D1272934

# CORK WARS

# CORK WARS

## INTRIGUE AND INDUSTRY IN WORLD WAR II

David A. Taylor

Johns Hopkins University Press
Baltimore

Printed in the United States of America on acid-free paper
2 4 6 8 9 7 5 3 1

Johns Hopkins University Press
2715 North Charles Street
Baltimore, Maryland 21218-4363
www.press.jhu.edu

Library of Congress Cataloging-in-Publication Data

Names: Taylor, David A., 1961 July 9– author.
Title: Cork wars : intrigue and industry in World War II / David A. Taylor.
Description: Baltimore : Johns Hopkins University Press, 2018. |
Includes bibliographical references and index.
Identifiers: LCCN 2018010714 | ISBN 9781421426914 (hardcover : alk. paper) |
ISBN 9781421426921 (electronic) | ISBN 1421426919 (hardcover : alk. paper)
| ISBN 1421426927 (electronic)
Subjects: LCSH: Cork industry—History—20th century. | Strategic
materials—History—20th century. | National security—History—20th
century. | World War, 1939–1945—Economic aspects.
Classification: LCC HD9769.C73 T39 2018 | DDC 338.4/7674909044—dc23
LC record available at https://lccn.loc.gov/2018010714

A catalog record for this book is available from the British Library.

*Special discounts are available for bulk purchases of this book.
For more information, please contact Special Sales at 410-516-6936
or specialsales@press.jhu.edu.*

Johns Hopkins University Press uses environmentally friendly book
materials, including recycled text paper that is composed of at least 30
percent post-consumer waste, whenever possible.

# CONTENTS

## PART 3. BEYOND VICTORY

# MAIN CHARACTERS

## McManus

CHARLES McMANUS SR.
*Born in Baltimore, Maryland, 1881*
Inventor, CEO of Crown Cork and Seal

EVA OLT McMANUS
*Born in New York, 1890*
*Married in New York, 1909*

CHARLES McMANUS JR.
(elder son)
*Born in the Bronx, New York, 1914*

MARY SCHAFFER McMANUS
*Born in Maryland, 1916*
*Married 1938*

WALTER McMANUS (younger son)
*Born in New York, 1918*

## Marsa

MELCHOR MARSA SR.
*Born in Barcelona, Spain, 1883*
Came to America in 1906, worked for International Cork in New York, returned to Spain and Portugal in 1934 with Crown Cork and Seal

PILAR MIR MARSA
*Born in Barcelona, Spain, 1888*
*Married in New York, 1910*

GLORIA MARSA (younger daughter)
*Born in Brooklyn, New York, 1924*
Educated at Packer Collegiate, worked in business, married in late 1940s, relocated to Mexico

## DiCara

GIUSEPPE (JOE) DiCARA
*Born near Messina, Sicily, 1876*
Railroad worker and wine-maker
Arrived in Baltimore, 1905

ROSA CAVALLARO DiCARA
*Born near Messina, Sicily, 1890*
*Married 1905*
Arrived in Baltimore, 1905
Raised six children

FRANK DiCARA (youngest son)
*Born in Baltimore, 1926*
Teenage bomber factory worker, WWII veteran, later worked at Crown Cork and Seal

IRMA CASTAGNERA DICARA
*Born in Baltimore, 1929*
*Married in Baltimore, 1948*

## Additional Characters

HENNING PRENTIS JR.
*Born in St. Louis, Missouri, 1884*
CEO of Armstrong Cork

HERMAN GINSBURG
*Born in Lithuania/Russia, 1900*
*Married Bobbie Shapiro 1935*
Crown Cork International director

W. MICHAEL BLUMENTHAL
*Born in Germany, 1926*
Princeton economics professor,
Crown Cork International employee,
US treasury secretary (1977–79)

WOODBRIDGE METCALF
*Born in Grosse Pointe Farms, Michigan, 1888*
Forester at University of California, Berkeley

# AUTHOR'S NOTE

This book is narrative nonfiction, combining the fact-finding of journalism with literary techniques to create a dramatic story that is also true. All the characters are real people. The narrative draws from dozens of interviews and documentary research, including many government sources that were previously classified. I obtained some of those documents through Freedom of Information requests, others from personal collections. All are cited in the corresponding chapter's section in the Essay on Sources.

Passages that present a person's thoughts and feelings have been fact-checked with that person or his or her family, wherever possible. Dialogue appears as eyewitnesses to the events relayed it, or as it appeared in written or published accounts of the time. The only exception is a hospital scene where I added three words of dialogue.

In a few cases I have added details of the surroundings and everyday behaviors as they very likely happened, based on historical or other documentation of the period and that person's circumstances, for example, how the streets appeared as that individual took a particular route to the office.

# CORK WARS

Prologue

# THE BLAZE

· 1940 ·

> Nine acres of baled cork roared into flames yesterday afternoon. . . . The fifteen-alarm blaze . . . was fanned to a glowing inferno of smoke and heat under a northerly wind.
>
> *BALTIMORE SUN*, September 18, 1940

> Cork . . . has become of double interest because of its shrouded mystery, which has never been pierced to the extent of giving the world a complete and comprehensive story.
>
> WILLIAM BOYD, "Cork," 1992

On a late September afternoon in 1940, Charles McManus was looking forward to dinner with his wife, Mary, as he prepared to leave his office at the Crown Cork and Seal factory in the Highlandtown neighborhood of East Baltimore. The phone rang.

Crown Cork and Seal was the world's leading maker of "crowns," or caps for soda and beer bottles. "Cork and seal" referred to the thin slivers of cork inserted in the caps to seal them tightly to the bottle. Something as small as a bottle cap made Baltimore one of the world's leading importers of cork. In the age before plastics, cork was *the* word of the future. Around 1900 an inventing boom capitalized on cork's unique abilities to insulate and form a tight seal. One key invention was a process for breaking up cork fibers and laying them out in thin layers, like a flexible version of plywood. "Composition cork" was perfect for the industrial age. Cork already had a notable history across Europe for its adaptability to different tasks, including stoppering barrels. Now it found a niche in American industry with even greater malleability of form. Composition cork's pliability met manufacturers' needs, from the automobile industry to creative new flooring businesses. Besides bottle caps, Crown Cork and Seal churned out oil-tight cork gaskets for the auto industry and the growing aircraft business. Cork held a special place in the future depicted at the World's Fair in New York in 1939–40.

Yet cork's roots clung firmly to the past. Companies like Crown Cork chartered freighters to load harvests from the oak forests that rimmed the Mediterranean in Spain, Portugal, Morocco, and Algeria. The cork oak tree was unique in that it grew an outer layer of bark that could be peeled off the trunk and branches without killing the tree, like shearing wool from sheep. That outer layer grew back, and workers could harvest the forgiving tree again in eight or ten years. Cork is unique in the genus *Quercus* in another way: it is the only evergreen oak. Otherwise, the tree's biology resembles other *Quercus* species: it produces acorns and both male and female flowers. Cork forests shelter a surprisingly diverse mosaic of wildlife and overlapping communities of plants and animals, with up to one hundred species of flowering plants coexisting in the space of just one quarter-acre.

Rural families residing near cork forests had made a living from the trees for many centuries, going back to ancient Rome. Over time, an industry emerged that gathered the lightweight sheets of cork from the forest, steamed them flat and lashed them together in bales, then hauled them to ports in Seville, Lisbon, and Barcelona. The freighters left Spain and Portugal crammed high with bales of the stuff. Even full, the ships sat high in the water as if they were empty, like ghost ships.

Across the ocean, the freighters unloaded in Baltimore and New York. These ports had complex ecosystems of their own, human ecologies that churned with their own dynamics. A city's landscape held vast stretches of pavement, steel, and humanity that thrived and went fallow in a rhythm driven by the ebb and flow of opportunity.

September was late in the cork harvest season, so the stockyard in Baltimore bulged with a year's supply—more stock than usual. With war in Europe and Nazi U-boats harassing merchant ships, Crown Cork, run by Charles's father, had bought as much as it could that season. The bales of cork rose in pyramids over sixty feet tall, as high as the Sphinx in Egypt.

His father was away on business, so Charles picked up the phone. When the man yelled, "There's a fire in the factory yard," he felt the words in the pit of his stomach. By the time he got close to the stockyard, it looked like the sky was melting.

Charles had heard the stories from his father. Twenty years before, a cork company near Boston had gone up in a big factory fire. The business never recovered. Crown's chief competitor, Armstrong Cork, had also suffered a massive fire in Pittsburgh in the early 1900s. Charles had heard his older colleague, Melchor Marsa, talk about a plant fire in Brooklyn in the 1910s.

Now this one, *his* fire, was the biggest of all. The newspapers would later say that you could see its glow from Philadelphia and from Annapolis—a rockets' red glare on the horizon. It was the biggest fire anybody in the city could remember. Families sat out on their stoops and stared up at the clouds like embers—no one alive had ever seen the night sky as light as day. The streets looked like a theater set. At the railway sidings off O'Donnell Street, people climbed aboard flatcars as though they were grandstands for the inferno show.

Charles stood in his shirtsleeves, staring at the firestorm. His tie and collar, soaked with sweat from the rush and the September humidity, were suddenly baked dry within minutes, the heat was that intense.

No matter which side of the blaze you were on, a hot wind came from its direction. The structure of the cork itself unleashed more and more fuel; each cell was packed with air, like a tiny zeppelin. When a flame hit cork, the cell erupted and released more oxygen to ignite more cork. "You literally couldn't put it out because of the cellular structure," Charles explained later.

The cork burned all night. Trucks full of firefighters kept coming, but the fire kept raging. It burned all the next day. It burned for two full days, exhausting four hundred firefighters before they finally got the flames under control. The fire ravaged nine acres of the stockyard and destroyed a half-million dollars' worth of cork.

When daybreak came, Charles saw the scope of the disaster clearly. Sparks were still flying high and far, floating dangerously close to the Standard Oil tanks a few hundred yards away.

Charles wanted to break the news to his father carefully. But there was no chance of that—the Ohio plant manager who answered the phone grew hysterical immediately. The older McManus took the receiver and said he would fly back the next day.

In the following days, everyone asked, *Who started the fire?* People spoke of sabotage. The fire placed McManus's company on the government's radar. It had to be about more than just bottle caps.

* * *

For people who lived in Highlandtown, the fire and the question of its origin filled them with dread. Many families had someone who worked at the factory—someone whose job was now endangered. Many of these had immigrated from Italy, Greece, or Germany. A generation before, Highlandtown had been an outskirt known as Snake Hill, home to sausage factories and other messy businesses that city officials liked to keep out of sight.

At a house on Pratt Street, Frank DiCara, the youngest of six, had just turned thirteen. A slight boy with a fair complexion and limp black hair, Frank stood by his bed staring out the window that night, wondering about the source of the smoke and the strange glow. Occasionally he saw sparks shooting skyward. The scene stirred the kind of thrill and vague terror that he got from watching Boris Karloff movies at the Rivoli Cinema.

Frank's parents, like most of their neighbors, had come from Europe. Rosa and Giuseppe had met on the ship from Sicily. They both were from Messina, and both had relatives in Baltimore. They had no money but found a community in the congregation of Our Lady of Pompei on Claremont Street. Giuseppe took a job as a track laborer for the B&O Railroad, starting at a dollar a day. Scrimping, the couple had saved just enough to buy the house on Pratt Street, which had no heat or indoor plumbing.

With war looming in 1940, families like the DiCaras came under as much suspicion as German Americans and nearly as much as Japanese Americans, all of whom were tarred by the taint of the Axis Powers. That fall politicians facing congressional elections decried "fifth-column" insurgents. The FBI compiled a list of Italian Americans whom they felt should be rounded up when America entered the war. They would be the ones to pay the cost of Americans' fears.

Frank, who looked like he didn't get enough to eat, endured all the names: *dago*, *wop*, *guinea*, *lowlife*. He was heading for high school and a different life from that of his parents. He knew that much. But that churning glow outside his window unsettled him. It simmered all night like a volcano. The next night it was still there.

* * *

Months after the Baltimore disaster, a fire erupted in New York Harbor aboard a ship docked at Pier 27. It consumed almost 300,000 pounds of the ship's cargo, cork.

The Marsa family lived across the East River in Brooklyn. Melchor and Pilar Marsa had both migrated from the Catalan region of Spain when they were young. Melchor had worked for a competitor of Crown Cork, until a legal battle and the crash of 1929 wiped out the savings the couple had built over the previous two dozen years. Melchor went to work for Crown Cork and Seal, managing its plants in Spain and Portugal. He didn't know it yet, but his life experiences made him a prime candidate for recruitment as a spy.

His Crown Cork job in Europe thrilled his youngest, a bright girl named Gloria. She was a gift in his old age, he joked. The fourteen-year-old was good with languages, she relished reading as well as stylish hats, and she loved travel.

Overseas, the family grew close, the way many expatriate families do. They spoke Catalan at home, relishing its precise and classical rhythms. Through her father, who was forceful, respected, and a man of his word, Gloria glimpsed a world of honor and style. She tagged along on his business dinners and visited his office in Lisbon. Gloria imagined herself as a foreign correspondent. But the cork fire in New York Harbor signaled a flashpoint that would pull her family into the war engulfing Europe.

* * *

The McManuses, Marsas, and DiCaras gravitated to the cork industry from different starting places. For these three families—owners, managers, and workers from three waves of immigration—cork proved a dangerous connection during the Second World War. Each family would be tested—one in the war industry and its profits, one drawn into a web of spies and its dangers, and one plunged into the frontlines.

Their stories were connected by the unique properties of a seemingly innocuous substance. Pliability and resilience were fundamental to cork's physiology. The air in cork cells (like the *O* at the center of *cork*, bottled by consonants on either end) allowed the cork to compress and seal the space between two objects: between a glass bottle and its pressed metal top, or between a sandal and the sole of your foot. That sealant quality would, during the war, tie cork forests to a secret mission in North Africa that whisked a planeload of acorns from Morocco for plantings designed to free America from dependence on foreign materials. America's national security machinery in wartime manufacturing would pull together disparate communities—from 4-H gardeners growing cork seedlings in South Carolina, Louisiana, Arizona, and California, to secret intelligence agents working undercover in the industry, to factory workers in Baltimore and Detroit—into a large mosaic.

This narrative of the bottle-cap industry also opens a window on the turbulent social dynamics of wartime. The story shows how quickly national security scenarios can change lives, and how deeply these situations become enmeshed with commerce, immigration, and the environment. Families who have been Americans for a generation see their patriotism challenged. Young second-generation Americans who stumble into trouble either regain their footing or fall into poverty. The story that unfolds from the 1940 Crown Cork blaze affected people entwined with cork oak's fate long after the war ended, into the nuclear age.

Today, with the world turned toward synthetics, large cork oaks—rugged, woolly-looking giants from an older world—are a rare sight in America. In the savannah slopes of Spain and Portugal where they are native, the oaks spread their canopies over large swaths of the hills. A passenger on the train from Alenteju, the Portuguese heartland, west to Lisbon passes through airy groves of cork oaks. Outside the train window, the trees emerge from a morning mist like ancient warriors. How did these brooding creatures create a web of industry, espionage, and mayhem? The answer lies with people, of course.

# PART 1

# FROM BOTTLE CAPS TO BOMBERS

Chapter One

# McMANUS PEELS THE APPLE

## · McManus: 1936–1939 ·

This story is completely different than what you'd suspect.

CHARLES McMANUS JR.

Charles McManus Jr. was in his last year at the University of Maryland in 1936, when the *Baltimore Sun* published a feature, "The Story of Cork: From Tree to Bottle," with photos of workers stripping cork trees in Spain. Crown Cork and Seal and its main rival, Pennsylvania-based Armstrong Cork, were marketing cork as a modern material not just for bottle caps but for floors and roads and automobile dashboards. Soon Frank Lloyd Wright, with his masterpiece Fallingwater in Pennsylvania, was making cork chic for flooring and walls.

*Baltimore* magazine published another feature, a profile of Charles McManus Sr. and Crown Cork and Seal. The article started with a description of a bottle cap: "A small thing . . . a bit of tin fashioned around a thin wafer of cork," it began, "a useful, inexpensive article, invented in Baltimore and found everywhere."

America had few business schools or incubators in the mid-1930s, and the country's first MBA program, at Harvard, was not yet thirty years old. Still flattened by the 1929 crash, Americans had little faith that business was the future. Compared to their twenty-first-century counterparts, average Americans then held practically socialistic views about what the government should provide citizens and the role of business. Polls showed that Americans tended to be fairly optimistic: about half of the population expected business conditions to improve in the next six months. Enterprise was still at the heart of American identity; people saw business as important although severely challenged by the

Depression. But the chasm between big corporations and the man in the street was wider than ever.

Charles McManus Jr. lived in that chasm. He considered himself a kid from the Bronx, where he was born, and he sounded like one when he spoke. He was twenty-two years old, tall, gangly, and deaf in one ear from a childhood infection. He was coming to terms with having a father who was one of the richest men in Maryland at a time when most families were still knocked flat. Crown Cork and Seal was the city's ambassador to the world, the *Baltimore* magazine article said. The company had subsidiaries in Canada, England, France, Belgium, Holland, Spain, Portugal, Algeria, and Brazil.

Crown's factory in East Baltimore, according to the article, was "the largest manufacturing plant of its kind in the world." It sprawled over seventeen acres and created everything necessary for making bottle caps—the tin, and the litho press for colorful brands—everything except the cork, which arrived by rail from the harbor, almost visible from the factory yard. The plant had grown in spurts over decades, evolving with the vision of Charles Sr., who was also an inventor. One building housed what he called "the breaker room," with machinery that he had developed for granulating cork slabs and pressing them into thin, link-sausage cylinders of composition cork. Another building contained the machine for slicing the cylinders into disks.

Over the decades Baltimore itself had grown, shrunk, and grown again, layered with the strata of sugar refineries, metal and chemical plants, meatpacking plants, canneries for seafood and spices, and builders of ships and airplanes. By 1940 it was the sixth largest port in the world. Guidebooks described the city as "charmingly picturesque in its ugliness," with block after block of redbrick row houses, crooked alleys, and "lordly mansions." It was marked by ethnic divisions and rigid segregation, too, with African Americans forced by housing and work restrictions into one neighborhood in the city's northwest section. Driving across town, Charles Jr. saw how the city layout tortured the compass, with West North Avenue and East North Avenue, East West Street and West West Street, and even a Charles Street Avenue Boulevard Road.

Charles stood at the factory gate and felt both a part of its ramshackle industrial glory and at the same time outside it. The workers walking past him to clock in knew he was the owner's son. Some of the staff were actually family, such as his mother's younger brother Leonard, who worked across several divisions of the company, and his father's brother who worked in the litho department. Charles had known others as long as he could remember; they might as well have been blood relations. It was a strange feeling, familiar yet not exactly warm.

His father was always thinking up new opportunities for the company. At home the old man carved small models for new contraptions and arranged them on the family pool table. He inspected the configurations from every angle, rearranged them, and considered the flow of materials through the factory. When Charles Sr. settled on the layout that integrated the new elements best, he would bring in an architect.

The facility expanded rapidly as demand for new products grew. The company added operations for the automotive industry that produced cork mats, gaskets, windshield glazing strips, sun visors, interior trim panels, fender wells, and antisqueak devices. McManus's composition cork was a good fit for other industries too: railroads used cork "sound and vibration pads"; the shoe industry used cork "box-toe material," as well as leather substitutes. Crown made cork and rubber sheeting, cork heel pads, even monthly calendars made from cork sheets sliced thin as paper. The McManus era had transformed the bottle-cap business.

Charles McManus Sr. was an unlikely mogul, born in 1881 in a rough neighborhood near Baltimore's Penn Station, the son of a small-time carpenter and the grandson of Irish immigrants. One uncle managed a bar. Another uncle, a cop, once collared a juvenile delinquent named George Ruth, later known as "Babe." Baltimore's streets were violent. One day a boy brought a gun to school. It went off and the bullet struck eleven-year-old Charles in the face, leaving him horribly wounded.

He was lucky to survive. Shrapnel lodged behind his left eye, making him partially blind. A slow, painful recovery forced him to miss so much school that he eventually dropped out.

As a teenager in the 1890s, he took a job in an umbrella factory and then in steel mills and bars in eastern Pennsylvania. Tending bar in Doylestown one day, he heard a customer complain about a bad bottle cap. "If somebody could make a cork that doesn't fall apart," the man said, "they'd make a million dollars." That galvanized the young man's attention. A stronger bottle top became his holy grail.

His rise in the bottle-cap industry was not as simple as the magazine article indicated. Crown Cork's founder, William Painter, had grown the company from a small shop on Monument Avenue. He was the kind of inventor that the young McManus dreamed of becoming. Painter founded an electric railway, created a counterfeit coin detector, and developed a pump for drawing water out of sunken ships in Cuba after the Spanish-American War. A Baltimore native, Painter came from a solidly middle-class family. He had grown up on a farm and apprenticed with his uncle's company in Delaware, which manufactured patent leather goods. (Another uncle was the illustrator Howard Pyle.) Painter took a job as a mechanical engineer with a machine shop back in Baltimore, but his mind was always probing new gadgets: multiplex telegraphs, telephonic instruments, hydraulic pumps, or lamps that burned coal oil. When an idea came to him, he took a piece of chalk and sketched a diagram of his notion on the shop floor. In August 1891, after a visit to Narragansett Pier in Rhode Island, he made a drawing of a new bottling method, which he believed would revolutionize the industry. He filed his patent for a bottle top sealed with cork in early 1892 and launched Crown Cork and Seal.

His timing was perfect. The following year an accident on the docks of the East River revealed a transformative piece of cork's nature. John T. Smith owned a boat works business on lower South Street on the East River. In a cast-iron kettle, Smith and a few workers steamed oak to shape them for boat frames. His business also made life preservers and buoys by packing broken-up cork into canvas life jackets, scooping the cork grains into tin cylinders. One day a cylinder clogged with cork rolled into the embers of the firebox under the kettle. The next morning while cleaning out the firebox, Smith found the tin and noticed that the cork fragments inside had not burned up—instead they had fused into a chocolate-brown mass. Intrigued, Smith set the blob aside. He created the conditions again the next day and discovered what he called "Smith's Consolidated Cork": a material that could be used as an insulator and sealant. That same year, 1893, Smith sold his patent to a company called Stone and Duryee in Brooklyn, and it began manufacturing fused cork insulation for steam pipes.

With that happy accident, cork disks became more effective seals in bottle caps, and Painter's company grew. Bottled drinks including beer were becoming popular. In three years he automated the process of filling and capping bottles with a foot-powered machine. By 1898 Painter was the toast of Baltimore, celebrated with a banquet by the American Bottlers Protective Association at the grand Music Hall. He bought a big new house on North Calvert Street in the heart of the society district. But he didn't get to enjoy his success for long. Painter's health declined as his fortunes rose, and he retired from Crown Cork in January 1903, leaving the company in the hands of his family. It coasted for two decades.

Meanwhile other industries flourished. After the First World War, many emerging businesses capitalized on cork's essential flexibility. There was a blossoming of inventions that exploited cork's compressibility, its sealant quality, and its resilient plasticity. The trade in cork imported from Europe and North Africa grew prodigiously. This was when McManus Sr.'s resilience came to the fore. Leaving his job as a Doylestown bartender, he went to New York, which was then the Silicon Valley of invention. He signed up for night courses in chemistry and engineering and looked for cheap lodging.

He rented a room in Manhattan from a couple named Furnas. Mr. Furnas, a housepainter, and his wife had no children; McManus was "their son who showed up," he would say. The couple encouraged him as he tested recipes for a better bottle cap at the house—even when his experiment clogged the pipes in the kitchen. In fact the plumbing disaster was a technical success: it proved his new material could make a watertight seal on the inside of a cap. McManus was desperate to make something new. When he got his first patent, for a new type of jar cap, and a contract to produce it, he and Mr. Furnas made the Gulden mustard caps in the basement by hand.

Prohibition dealt a severe blow to brewers, which dominoed into disaster for bottlers. It initially hurt McManus too, just as he was trying to grow his business, New Process Cork. Friends advised him to find a different industry, but McManus doubled down and expanded from beer into the soft-drink market—which turned out to be the only part of the bottling business that grew. The soft-drink industry also benefited from innovative marketing by Nehi, Coca-Cola, and several near beers. Meanwhile his main rival, Crown Cork, was feeling the pinch.

McManus was on his way to becoming a major player in the industry of "crowns," or bottle caps. Then, on one of the first commercial flights from the United States to Havana in the early 1920s, he happened into a conversation with a passenger named Bill Meeker, who was married to a granddaughter of William Painter. As the two men talked, Meeker confided that the Painter family wasn't getting the income from Crown Cork that it used to and that the company seemed to be sinking. McManus said, trying to sound offhand, "Well if you can get me all of the Crown Cork stock, I'll try to make a deal."

Thus in 1927 McManus "astonished the crown industry," according to newspaper accounts, by acquiring Crown Cork and Seal. He returned to his hometown as an industry leader, the carpenter's son made good. After twenty years away, McManus came back to a growing city. Baltimore had a wide range of housing options for working-class families, thanks in part to builders like his grandfather. The city also had a lot of low-income housing, such as the neighborhood around Crown Cork's Highlandtown plant, where he brought much of the equipment from his New York operation.

The woman who accompanied him was less enthusiastic about their new home. Eva Olt was a New Yorker who had met McManus on the train to Albany. She had an independent, generous personality and was a cosmopolitan first-generation American. Her father was a furniture maker from Germany, her mother was from Denmark. Eva loved to read and travel and kept a diary throughout her life. McManus introduced himself on the train as "an experimenter." They got on well, and the couple married in 1909 at the Little Church Around the Corner in lower Manhattan.

Now they traveled the world together first-class on the most elite ocean liners. In their passport photo, McManus looked like a quintessential middle-aged businessman: sober suit and bowtie, hard features, thinning hair. Beside him Eva sported a tailored dress and a hairdo more stylish than was worn by most women of her time.

"The story of cork," the *Baltimore* magazine article concluded, "is a most interesting and unusual narrative, filled with color and romance."

Romance for Charles Jr. had come in a low-key way, after a painful series of doctors' visits as a teenager. His early struggle with deafness had involved many rounds of appointments with specialists, to little avail. For a while, he was completely deaf, and no treatment had much effect. Finally, doctors at Johns Hopkins managed to restore hearing in one ear.

Throughout the ordeal at Johns Hopkins, he stayed with cousins downtown, where he met Mary Schaffer, who lived next door. She was fourteen, he was sixteen. They began seeing more of each other. Charles learned how to drive mainly so that he could make his way from his home on the city's northern edge to see her. The doctors' visits also continued, about once a month.

As a young man, Charles believed that his bad ear had led to the best thing in his life: Mary. His mother didn't see it that way—she saw

Mary Schaffer's father as just a plumber whose daughter was not worthy of Eva's son. But by Charles's last year in college, the relationship had grown serious.

One day in the fall semester of that year, Charles Jr. was in the library. As a diversion from studying, he looked up cork and stumbled upon a surprisingly in-depth article by Dr. Giles Cooke, a professor there at the University of Maryland. When Charles was home for Thanksgiving break, he mentioned this discovery at the dinner table.

His father said, "Oh, I just hired him."

Charles also learned his father had a sideline in agriculture. The old man had bought a poultry farm at Willow Mill, farther west in Maryland, where he consulted with Dr. Cooke on new binding materials that could be used in bottle caps. Egg whites had provided a natural commercial sealant for generations (for instance, luthiers used egg white to bind violin parts before varnishing), and bottle-cap manufacturers used so much dried duck albumen (the protein in egg whites) that McManus needed to import duck eggs wholesale from China.

In the autumn of 1936, world events in both the eastern and western hemispheres were threatening to affect Crown Cork and Seal. News from China reported on the long-running civil war there, and in November Japanese-supported forces invaded the northeast. These distant events triggered his father's visits to the Willow Mill farm. To replace lost egg-white imports, McManus and Cooke were tinkering with different chicken breeds and feeds, searching for a substitute egg white.

Eventually their chickens produced albumen that, his father said, was virtually identical to the imported version from Chinese ducks. Not long after that, McManus ended his visits to the poultry farm. It had done its job as a testing lab. He donated the farm, along with a herd of cattle, to the University of Maryland as a research station. The total value of the gift was $1 million.

Young McManus couldn't say when, exactly, he decided to join his father's company. Growing up, it was simply an unspoken assumption between them. In 1937 he joined his father on a trip to Europe and North Africa, where they observed cork harvesters in Spain, Portugal, Morocco, and Algeria. It was Charles Jr.'s first time to the Mediterranean. In Spain, McManus had two plants: one in the south near Seville and one north of Barcelona.

Father and son arrived amid the eruptions of Spain's civil war. The

Spanish military had seized power from the elected Republican government. Junior found the whole experience of Spain confusing and fascinating. In Seville they met Crown Cork's manager for Spain, Melchor Marsa, and his family. In Portugal, four years into the stern regime of António de Oliveira Salazar and his Estado Novo government, the McManuses ventured outside Lisbon to where cork was harvested. As they drove out of the city, modern life in Lisbon gave way to wooden homes, chickens, and horse carts. Up a narrow road to a farm collective, and off to the side of the house, lay a big open yard. Dairy farmers, shepherds, and others who kept a sideline in cork gathered the cork harvests.

After some small talk with the farmers, the McManuses walked to where the crew—a dozen men and a few women—laid out big sheets of cork and, like sculptors, started cutting them into large rectangles with savagely sharp knives. The scene had the air of an ancient pagan ceremony. The carvers, in their wool jackets and trousers, were attacking great sheets of thick cork with picks and saws and an antlike intensity of purpose. Each slice—first a vertical slice on the trunk, then horizontal cuts for peeling—seemed preordained by the grain and branch patterns. They made the slices very carefully, coaxing the cork from the trees. Then they wedged the pointed handle of an ax or machete into the slit. After a few minutes of jimmying, they pried the cork from the trunk. It popped right off.

The workers sorted the slices into stacks for particleboard and insulation, and into piles for wine stoppers. Here was Crown Cork's natural source in all its variety. Charles watched the platoon of workers expertly slice one cork sheath in a huge ring from the trunk. They peeled the tree so perfectly the slab could stand up by itself, retaining the shape of the absent tree trunk. The scene of those tremendous ghostly tree husks pushed together, the empty trunks of cork standing on their own, amazed him.

Under Portuguese law, all cork oaks belonged to the government and officials regulated the harvests. The rule was that you could harvest the bark every seven years. On a young tree you could peel only the lower trunk. As the tree aged, the harvest could go out onto the branches, but the cork quality decreased. The better, thicker cork was down around the tree's base.

After the peeling and sorting, Charles Jr. watched the harvesters haul the great husks to the collective's courtyard, where they steamed the slabs, flattened them out further, and sliced and lashed them in giant bales. A few workers snipped out the knots and bits with irregular grain, like surgeons. They saved the better quality cork—about two inches thick—for the wine stopper buyers. The platoon of young men and women with blades prowled the yard.

This episode in the countryside struck Charles Jr. profoundly. The harvest workers made him think of his grandfather's carpentry crew, assembling house frames, a scene both foreign and familiar.

* * *

Soon after that trip, Charles was working in his family's dairy operation outside Baltimore. One day his tasks brought him to the Crown Cork factory, where the manager of the machinery division looked him in the eye and said, "Come in here and learn machine work."

Working downtown was eye-opening. In navigating the city and daily life, Charles learned how to cope with his father's wealth and generosity. At Penn Station, the porters all knew Mr. McManus by name. Every time the head porter carried the old man's suitcase, he got a twenty-dollar tip, even in the depths of the Depression! Charles Jr. had a father that no waitress ever forgot because the old man would put down a twenty-dollar tip on a two-dollar check.

The trouble came when he returned to the restaurant without his father. The waitress paid extra attention to his table, expecting a huge tip. But he didn't have the cash that the old man did. The waitress asked about "that man you came with last time." Who was he? How was he doing?

The truth—that Charles didn't have as much money as his father—was never convincing. So Charles and his brother tried various stories—none of them entirely satisfactory. The only one that worked was the "rich uncle" story. Saying that the man was his rich uncle was a tactic that Charles used again and again.

Besides family members like his uncle Leonard, the Crown Cork business also drew workers from his father's first company in Brooklyn, New Process Cork, men who had been with McManus for more than twenty years. His father could be generous with opportunity. Herman Ginsburg had arrived in Baltimore as a teenager in 1919 and got a job in the Crown mail room. Herman had no high school diploma, but he studied after work and got a law degree from the night school at the University of Maryland. Within six years, he rose to management and became director of Crown's international operations. Ginsburg was perceptive and adept at negotiation. Seeing an idiosyncratic and ambitious spirit like his own, McManus established the young man in a small office in Jersey City and gave him a surprising degree of autonomy in managing the overseas plants and sourcing the raw material.

The country was finally climbing out of the Depression in 1939, when the future arrived in the form of the New York World's Fair. Going to the fair in Queens was practically an obligation for anyone who had dreams and ideals. During one of Charles Jr.'s trips to Manhattan he crossed the East River, paid his seventy-five-cent admission, and walked past the dozens of cafés, bars, luncheonettes, and oyster and chophouses that catered to the millions who came to the fair.

At one exhibit, McManus watched machines make nylon stockings, spin acetate yarn, and mold Lucite plastics. In Kodak's Great Hall of Color, Kodachrome images flashed on nearly a dozen giant screens. A forty-foot waterfall gushed outside the electric utilities exhibit hall. Technicolor movies and animated puppets described how petroleum flowed from the American Southwest through pipelines to refineries.

At an exhibit called Futurama, crowds ascended a circular ramp to where an attendant helped each person into a gliding wingback easy chair. As the ride began, each passenger listened to a narrative coming through a built-in speaker at his or her ear. The chairs circled on a conveyor as the narrator explained the features of the exhibit below ("the city of 1960, with its abundant sunshine, fresh air, fine green park ways—all the result of thoughtful planning and design"). As they left Futurama, visitors passed through the Communications Zone, which showed how in twenty years they would be receiving world news around the clock, at home, on a noisy teletype machine.

Charles Jr. made his way to the Ford exhibit, which had a sleek, all-glass facade and a huge turntable where more than one hundred animated puppets reenacted the process of making the copper, steel, and cork parts used in Ford cars. The exhibit was a machine geek's dream. Its Road of Tomorrow ride took people along a life-size smooth elevated highway, complete with spirals, cloverleafs, and overpasses, cushioned by a surface made of cork and rubber. (Ford later added a wing to the exhibit, acknowledging the past with a horse's rebuttal to the automobile age in a ballet titled *A Thousand Times Neigh*.)

Charles was surprised how the exhibit made him see Crown Cork in a new way. Wandering through the fair gave him a childlike rush of excitement. The international area hosted pavilions and exhibits from fifty-eight countries, including prominent spaces for Great Britain, Italy, and the Soviet Union, as well as the League of Nations. After September 1939, the Polish participants at the fair draped their exhibit space in black, a reminder of Germany's invasion of Poland that month. A giant sculpture of Peace at the entrance of the Court of Power suggested another age.

The fair's cork-paved highway, the smooth-riding easy chairs, the technology-studded landscape that was uniformly prosperous—all these belonged to a world far from the events that roiled the globe in 1940. Later many would look back on what the fair got wrong about the future. At the time, though, the fair's planners presented, against a bleak backdrop of current affairs, a World of Tomorrow that blended consumerism with shared ideals.

* * *

Charles Jr. had worked for the company for two years when, in a move that defied good sense, his father decided to transfer him to England to help manage Crown Cork's troubled British subsidiary. Junior had just married Mary Shaffer when he received the cable from his father in Europe: there were two reserved spots on the *Queen Mary* for the newlyweds' passage to England. "Get over here," the cable said.

Not sure what lay ahead, the couple put their belongings into storage and set sail. For this assignment, Charles Jr. would work in Herman Ginsburg's international division, but he would have little direct supervision. The England plant made bottle caps and shipped cork disks to Crown's European plants in Belgium and the Netherlands.

Only when Charles got to London did he learn the situation: the head of Crown's English operation, a man named Wallace, had spirited away a third of the plant's machinery—simply moved it to another location—not with a plan for embezzling but from fear of it being destroyed by German bombs when the air raids started.

Charles arrived at the plant and visited the factory floor, his footsteps echoing in the empty space. He asked the assistant factory manager, "What do you need to get production back up to normal?"

"Need to get the machinery back," the manager replied. That became Charles Jr.'s first task. He tracked down the equipment's new location and arranged for its return to the factory.

His assignment also threw him into the tangles of the customs bureaucracy. One afternoon Junior got a phone call from the Netherlands office of Crown Cork, asking, "Where's the tinplate shipment?" He had no idea. For bureaucratic reasons, the Dutch factory could not import tinplate (the thin steel coated in tin used in soft drink cans) from next door in Belgium—it had to come from England. But it took some digging to find out that British customs agents had held the shipment at the dock. A customs officer refused to let it out of England, saying it was a restricted material needed for the war effort.

Charles drove to the port and talked with the officer for an hour. He eventually found a loophole that consisted of a letter space: instead of "tinplate," he decided, they were shipping "tin plates." The shipment went through to the Netherlands.

Life in Europe got crazier for the young couple. Great Britain declared war on Germany in September 1939, and yet Charles and Mary stayed. To get some distance from the expected bombing raids on London, they moved out of the city center. Mary grappled with the mother tongue. At the grocery store, she couldn't get used to calling cookies *biscuits*, or remember what to say at the butcher counter. The car parts had different names too. But she made friends with other women whose husbands worked for Crown Cork, and she adapted.

Finally the danger grew too great. They boarded one of the last ships out of England before all commercial transports stopped indefinitely. Germany was sinking ships right and left, military or not; the *Athenia* had gone down off the Irish coast the previous month with more than 1,100 passengers aboard. The "U-boat peril," Winston Churchill later wrote in his memoirs, "was the only thing that ever really frightened me during the war." By cutting off Britain's lifeline across the Atlantic, Hitler could seal a British defeat.

The young couple sailed on the *American Trader*, on October 8. They stared out at the waves, unsure if they would make it home. Reports of more German-torpedoed passenger ships came over the wireless.

Months later, Germany began to rain bombs down in the Battle of Britain. Charles's education in world events was coming along.

* * *

With Europe unraveling, the American cork industry reeled as supplies became more and more unpredictable. Hitler invaded Czechoslovakia, Poland, the Low Countries, and France. Nothing assuaged his aggression. Spain was barely stabilized under Franco's repressive rule, and it was selling more and more cork to Germany.

In the countries that exported it, cork's subculture had vulnerabilities that McManus Sr. saw firsthand, including inequalities between the rural families who harvested the cork and the urban elites who owned the companies. Plus, cork supply chains were right in the path of war.

McManus did what he usually did when forecasts looked bad. He expanded. He bought companies in Illinois and California and began talks about possible new joint ventures with fellow Baltimore businessman and aviation legend Glenn L. Martin (whose company would eventually become the aerospace giant Martin Marietta). Martin had contracts with the US government to build more aircraft. Those contracts could mean new markets for Crown Cork's materials.

The federal government was creating incentives for companies to help achieve the titanic goals that FDR set for industry as the war effort loomed, recognizing that manufacturers needed more encouragement than just patriotic fervor. "If you are trying to go to war, or to prepare for war, in a capitalist country," said Henry Stimson, the secretary of war, "you have got to let businesses make money out of the process."

Crown Cork and Seal quickly scaled up, in the words of historian Robert Brugger, from stamping out bottle caps to making nine-thousand-pound gear rings for merchant ships and nose caps for armor-piercing ammunition. The company reported to stockholders that in 1941, sales soared to $45.09 million, 36 percent over the previous year. A government order in February 1942 forced it to discontinue "nonessential" tin can operations (for beer, coffee, motor oil, dog food, and so on), blocking a key commercial line of its products. Many other companies had been forced to shut down due to similar restrictions and rising costs. Many went into the war industry to stay afloat financially.

McManus navigated these currents, hoping to avoid political shoals. Other industry leaders, including Pennsylvania-based Armstrong Cork, objected strongly to the government restrictions. Armstrong's CEO, a prominent Republican with the blueblood name of Henning Prentis Jr., gave political speeches warning of impending doom to individual liberty and civilization if America continued on President Franklin D. Roosevelt's path of meddling in the economy. Prentis also spoke out against America's drift toward war.

Armstrong Cork (known today as Armstrong World Industries, a leading manufacturer of flooring products) had a longer history than Crown Cork and Seal, stretching back to 1860. Armstrong started as a cork-stopper shop in Pittsburgh, and provided bottle stoppers to the Union Army during the Civil War. The company expanded distribution nationally after Appomattox, and in the late 1870s it extended its reach overseas, importing corkwood and cork directly from Spain. By the 1890s Armstrong was the world's largest cork company, with more than 750 employees. It expanded again into flooring products, including linoleum.

Armstrong established a reputation for loyalty to its employees through benefits like health and dental care, which were rare in the early 1900s; the company was among the first to pay overtime wages. Management started offering staff paid vacations in 1924 and life insurance in 1931. Its philosophy was that a company should provide for its employees voluntarily, not because of government regulation.

Prentis had followed an academic's path into business. The son of a teacher, he was organized, articulate, a practiced public speaker, and a strong advocate for the "talented minority," by which he often meant people with education. Prentis read voraciously, making notes in the margins of his books, and was an avid photographer and sometime painter. He was the only cork executive with a master's degree in economics. As a young man, he took a job with Armstrong writing advertising copy, intending it to be a short stint on the way to a career in teaching, but he became intrigued by public relations. He became head of Armstrong's small advertising staff in 1911 and persuaded management to fund a three-year ad campaign with the *Saturday Evening Post*. As CEO, Prentis led the company's move to Lancaster, Pennsylvania, and diversification into rubber- and asphalt-tile manufacturing. Armstrong recovered from losses early in the Depression and reached record profits in 1936. Its stock soared.

Speaking for many American businessmen, Prentis in May 1940 urged the US Chamber of Commerce at its annual convention, "Let business set its own house in order." He asked the federal government to do the same. In Boston that fall he warned against getting involved in "Europe's war," telling industry leaders that domestic issues should come first and that national defense suffered when social programs took the foreground. In December 1940, after Western Europe had fallen and Greece and Hungary had joined the Axis Powers, Prentis insisted, "American industry is fundamentally opposed to war." He went so far as to lump liberals together with communists and Nazis: "Socialists, Nazis, Fascists, new liberals," he said, "are blood brothers under the skin. All of them deny the sanctity of the individual soul and hence all of them find little or nothing abhorrent in the idea of the all-powerful state."

This was close to the rhetoric used by the isolationist America First movement. Begun as the America First Committee in September 1940, the group launched a petition demanding that the United States stay out of the war. America First distrusted FDR and opposed his support for Britain's defense in convoys of ships and lend-lease aid. America First members held that American democracy could survive only by avoiding Europe's war. The movement attracted ideologues like Charles Lindbergh, who drew thousands to the cause with his speeches.

* * *

Cork oak forests around the Mediterranean were as seasonally predictable as apples or dates. Each year local harvesters sheared a different grove, like fields on a fallow rotation. In the cork groves designated for that year, men and boys brandished sharp, harpoonlike saws and peeled the thick, spongy bark from the trunks, leaving them looking raw, pink, and gangly.

America, the world's largest consumer of cork, relied almost completely on these sources, soon to be isolated by a Nazi blockade. Cork imports to the United States had soared to nearly 4 million pounds a month in August 1940. Two-thirds of that total came from Portugal. Many of the cork-laden freighters were headed for Baltimore and Crown Cork.

Cork had become a critical and irreplaceable material of modern life and war. It was essential for making everything from bomber plane gaskets and insulation to tanks, submarines, cartridge plugs, and bomb parts. McManus's composition cork made up a third of that total.

Charles Jr.'s father had assembled the largest corporation of its kind in the world. McManus held seventy-five patents that grew the reach of cork into everyday life. Amid the Great Depression's deep gloom, Crown Cork's sales volume had quadrupled from $17.5 million in 1929 to nearly $76 million. It likewise quadrupled the number of its employees. Armstrong Cork had grown nearly as fast. New products like aluminum cans made the industry even more modern. Profits were going up for the largest operators, and smaller fry were getting squeezed out. The government was bound to take notice.

Still, the yard at Crown Cork's Highlandtown plant, with its acres of high pyramids of rough bales, had an ancient quality. The cork arrived from the port by freight car and was moved into tall concrete silos. Workers in the cork yard piled the bales onto conveyors, each weighing about 175 pounds. The bales went to a breaker house, where machines crunched the cork slabs into small granules. It took two men to wrestle a bale onto a conveyor and two more at the other end to get it off. It was exhausting work, and the yard workers went home covered in black dust.

Charles Jr., back from England, was preoccupied as he walked through the cork yard that fateful day in September 1940, trying to stay on task. There was a lot happening that fall, and he didn't need one more thing to distract him. His father was visiting their plants in the Midwest, seeing how the company could adapt to a shift in demands if America entered the war. Charles didn't see how that would come about with the intense public resistance to war, but stranger things had happened.

He replayed that walk through the yard in his head, again and again, for months after the Highlandtown plant fire. Had he seen anything? He had noticed a few guys on break. Later Charles kept asking himself: Was there anyone he didn't recognize?

* * *

J. Edgar Hoover and his agents were busy. Earlier that year, the FBI began infiltrating a ring of Nazi spies based in New York. They were led by a man named Frederick "Fritz" Duquesne, a South African Boer who had lived in New York since the early 1900s. Duquesne had held an intense hatred of the British since his family's sufferings in the Boer War. With a shell business in Manhattan and orders from Berlin, Duquesne assembled a network of spies including Paul Bante, who obtained information about shipping targets and was preparing a fuse bomb. Another operative was Alfred Brokoff, who was born in Germany and had been a naturalized US citizen since 1929; he worked on the docks and tracked ships bound for England. Others worked as ship stewards and cooks, or took jobs for automobile makers. One of Duquesne's crew designed power plants for utility companies in New York. They all had received training in Germany and by the fall of 1940 were organized and mapping industrial sites along the Northeast corridor. The FBI had recordings in which Duquesne boasted of being an arson expert, with plans to blow up American facilities.

On October 17, one month after the fire, Ida Wingfield arrived at the FBI's Baltimore office and asked to speak to an agent about the Crown Cork disaster. She had observations that the factory blaze could have been an act of sabotage. Her account, recorded by Agent Lorton Chiles, also suggested possible links to other industrial fires on the East Coast. She described an encounter that might be a lead in the FBI's investigation. Wingfield worked as a housekeeper for a doctor named Leo Breit, who practiced medicine from his home on the southern edge of the city. She said Dr. Breit and his wife were both originally from Russia, and recently they had had foreign visitors and were behaving strangely.

Wingfield believed that Leo and Anna Breit, his wife, had some connection to the Baltimore fire, and possibly to another factory explosion that newspapers had reported at the Hercules Powder Company in Kenvill, New Jersey. Dr. Breit had very few patients but always seemed "to flourish in money," she told the agent, and "spend quite freely." Wingfield described the couple in detail: Leo, forty-seven, wore thick glasses with tortoise-shell frames, played piano, drove a blue Dodge Coupe, and "always traveled about" with two bulldogs; Anna, a few years younger, had long black hair, "penetrating" eyes, and "striking" features. Wingfield also described suspicious visitors to the home, people with foreign accents. One frequent visitor that fall, a woman, came to the Breits' home a few days before the New Jersey explosion, and then again on September 17, the day of the Crown Cork and Seal fire. She was "particularly excited" that day. Another, described as a "student," was approximately thirty-eight years old, with a heavy black mustache and glasses, and who, like the Breits, spoke with a Russian "brogue." Wingfield noticed the doctor had recently left a letter addressed to someone at the Glenn Martin Airport on his desk, and she described Dr. Breit's behavior that day of the fire as "agitated."

Wingfield would continue to work for the Breits, she said, and would watch for any further unusual behavior. The agent typed up and logged her account, adding, "There is no definite information that the subjects in instant case are actually involved in subversive activities." Still, the FBI pursued the lead seriously. The Baltimore office sent a copy of the file to Newark, where agents were investigating the Hercules Powder factory explosion. The file, marked SABOTAGE, suggested the FBI had reason to look for clues across the Atlantic.

## Chapter Two

# THE MARSAS RETURN TO SPAIN

### · Marsa: 1934–1939 ·

Eucalyptus trees are for us, pine trees for our children, and cork trees for our grandchildren.

PORTUGUESE SAYING

Ten-year-old Gloria Marsa was excited to be on an ocean liner, though her father had misgivings about their voyage. It was 1934, and they were on the *Excambion*, bound for Europe. They were leaving behind Brooklyn for a new life in Spain. Her father had taken a job with Crown Cork and Seal, as manager of the company's operations there. For Gloria, it was perfect.

Gloria watched her parents across the breakfast table. When her father accepted the job in Seville, he had proposed going by himself and letting the family stay in Brooklyn. "Europe is a tired continent," he told them. "It has old customs." From their home in Flatbush, Melchor and Pilar went to the opera, the children attended good schools, and they often got to see the Giants play at the Polo Grounds. Gloria's father cheered the team on to the National League pennant, one time cheering so hard that he strained a vocal cord.

The Marsas had visited Spain before to see relatives in Barcelona and the nearby coastal town of San Feliu de Guixols. Gloria's parents had come to America in their youth. Melchor Marsa's family had a livestock import business in Barcelona. His father had died when Melchor was still in school, leaving an older brother in charge of the enterprise, and Melchor gave up his studies to help with the company. He went out at dawn to inspect shipments of sheep at the docks and stayed at the office late at night poring over the books. But tensions emerged between the brothers, and Melchor increasingly felt shut out. Finally, he told their mother that there wasn't room in the firm for both brothers. His mother replied that Melchor's brother had a family to support; maybe Melchor should be the one to find opportunity elsewhere.

He arrived in New York in 1906, when he was twenty-three years old. On the immigration form, he listed his occupation as "merchant." After Ellis Island, Melchor stayed with an uncle in Brooklyn until he got on his feet. He tried a series of jobs: one in a restaurant, another at a spaghetti factory, where he soon became a manager.

The first decades of the twentieth century saw a surge in immigration from Spain. The number of arrivals nearly tripled after the first decade, brought on by rural poverty and crowded cities—the same factors that fueled immigration from across all of Europe then. The *New York Times* blamed local spikes in crime and the increase in anarchist activities on "the flood of Italians, Catalan, and Spanish propagandists who have reached our shores."

Pilar Mir, four years younger than Marsa, arrived from Catalonia with her family around the same time. Her father had died, and her mother's sister, who was already living in the United States, urged Pilar's mother to bring the girls. "I think they'll enjoy the country and I think it's wiser for them to come," the aunt said. Pilar, who had studied to become a teacher, fell in love with America and immersed herself in reading everything she could find to improve her English. She lived with her family on Rush Street in Brooklyn. Melchor lived across the bridge at 173 Pearl Street in lower Manhattan, near the harbor. They met in the Catalan community. Pilar was struck by the young man's seriousness, the way he carried himself, and his intense brown eyes. He courted her, and they married in late August 1910.

Eventually Melchor joined a company in Brooklyn, started by Pilar's cousins, for importing cork from their contacts in northern Spain. The Brooklyn factory had been leveled by a fire in 1908, but the company soon recovered and developed a niche in stoppers for pharmaceuticals. At International Cork, Melchor discovered he had an aptitude for making equipment work. His facility with business blossomed too. He retained his Spanish citizenship for almost two decades, even though he registered with the US Army during World War I.

Melchor made a name for himself in the business and became factory superintendent. Like Charles McManus, he was a self-taught inventor who authored a handful of patents. The first decades of the twentieth century were heady days for new materials and manufacturing. Marsa managed the company's production as it grew to national stature. In 1926 Crown Cork bought International Cork and absorbed it. Marsa sold his shares to McManus and retired from the cork business.

Then came the 1929 crash, devastating the family's savings. "When my father retired in the late 1920s," his daughter Gloria recalled later, "he invested in a variety of things—stocks, bonds and properties—so that there would be versatility. Should one thing fail, another could sustain you. Well, everything failed in '29."

With three children at home, Marsa needed to go back to work. But there was no work, and by the mid-1930s he had few choices. One neighbor with a produce business was so worried about the Marsas that he mortgaged his own house so that he could help if Melchor needed money.

But rather than accept assistance, Melchor approached his former rival at Crown Cork. Charles McManus was having troubles with his subsidiaries in Spain and Portugal. He needed someone who understood the dynamics there, to turn those three facilities around. So McManus offered Marsa the job in Spain, including a percentage of whatever profits the subsidiaries generated. "They're your babies," he told Marsa. "As long as they make money, you make money."

Returning home from the meeting, Marsa considered his limited options. All across the city, he saw how the Depression ravaged families. He was fifty years old; his chances for a fresh start were slim.

He and Pilar talked through alternatives. Melchor argued for going to Spain alone: the children would get a better education in Brooklyn; their lives wouldn't be disrupted. Was this really a time to take them to a country they had never known? But Pilar insisted that this was a challenge they faced as a family. In the silences between the two smart, stubborn people, the house stood still.

Brooklyn was home. Melchor didn't want to pull his family out of America. He had a keen sense of humor but a fierce temper also. Pilar had a temperament that matched her husband's. When people congratulated Melchor on his career, he would shrug it off. "Pilar is the one who gave me the support and the will to do, and to do more," he would say.

Thirty years after he had left Barcelona, Marsa was returning to Spain with his American family.

* * *

The crossing on the *Excambion* was rough. Each day the steward laid out newspapers in the ship's breakfast room with big headlines about another vessel sunk by the storms. The dining room was nearly empty—many passengers had no appetite and huddled sick in their cabins. Gloria chatted with her parents and her sister. Her brother had gone ahead of them to take a job with the company in northern Spain. When weather permitted, she stole a few minutes on the topside deck.

They landed in Gibraltar and spent the night there. The next morning they crossed the strait to the port of Algeciras, where a car picked them up and drove them to Seville.

Reaching Seville at dusk, Gloria looked out the window on the city's narrow streets, where cars vied with burros and peddlers on foot. She felt assaulted by strange sounds and smells. The whole city looked dim—was it lit by gas lamps? By the time they checked into the hotel, she felt as if she were being buried alive.

In the dining room, the emotions of the long voyage, the drive, and this strange old place tumbled down on her. She burst into tears.

Her father put his hand on her shoulder and sat her down on his lap. "Look," he said consolingly, "I know this is a big change. I know this is very different from life at home. But please, give it time. You'll acclimate and learn what this culture is like."

Spain, however, was pitching into chaos. After the leftist Spanish Popular Front won parliamentary elections in February 1936, the military rejected the result and mounted a coup. The military-led Nationalists moved to take the whole country.

Gloria soon came to know Seville better. She found charm in its shaded stone plazas, ancient streets, and *arboles de naranjo*. She was struck by the city's hidebound customs. Women rarely left home alone. When they did appear on the street, they were not allowed to carry anything. One time when Gloria's mother went to the grocery store to buy a quarter-pound bar of butter, the shopkeeper wouldn't let her carry it home. "We'll send it," he insisted.

The French School was located on a street so narrow that cars couldn't reach it. Gloria walked down the lane and at the end came to an imposing facade. It was probably only two stories high, but in that low-lying city, the school felt huge. The medieval entrance was wide enough for several horses to pass side by side. She thought, "I'm being sent to the dungeon."

Her classes were in French and Spanish. She didn't speak either one. (The family spoke Catalan and English at home. Her mother and father wrote their letters to each other in English.) Morning classes were in one language; afternoon classes in the other. The other students mocked Gloria's accent in both. What was worse, her classmates probably interpreted her slightly Catalan accent as a signal of sympathy for the Republican forces—sympathies she was mostly unaware of. In the confusion of a polarized Spain, she was at sea. Every morning and again after lunch, she made her way to the dungeon, bewildered and afraid.

One day in class, the teacher called on Gloria to read a paragraph from *Don Quixote*. She did the best she could with the sixteenth-century Spanish. She heard snickers from the back row. This time, the teacher stood up for her, saying, "Let me ask any one of you to read a paragraph in English and see if you do as well." From that day on, Gloria worked harder to earn the teacher's respect.

The news of the war from Madrid and Barcelona was frightening: horrifying stories of mass killings committed by both sides. Much later accounts would show the Nationalists to be especially brutal, but in the early months of the war, newspaper accounts of Republican atrocities, especially the murders of priests, shocked everyone.

* * *

Gloria's father worked to arrange the Seville office for Crown Cork and to manage its two Spanish factories, where initial processing prepared the cork sheets for export. Seville was the heart of cork history where, centuries before, Andalusian Muslims had taken up forest-based industries left by the Romans. Pliny the Elder had described the cork tree and its uses in his book *Naturalis Historia* in the first century CE. He wrote of its thick bark, which could be flattened into large sheets and was used to make anchor ropes, fishing nets, and stoppers for casks. Horace and other Romans also wrote about cork, struck by its usefulness for stopping wine casks and making beehives. Virgil's epic *Aeneid* mentions cork as a head covering for the soldiers of ancient Latium.

Muslim craftsmen in Seville had adopted the Roman techniques of making cork-soled shoes and had improved them. Granada had an enclave of cork shoemakers, and there was another near Madrid. The famously comfortable shoes became an export staple in the eighth and ninth centuries CE; their popularity spread across northern Africa and throughout the Muslim world, along with the name for the shoemakers—*qarrâqin*, from the Roman *corco*. For centuries, shoes were the primary use for cork all around the Mediterranean.

Not until the 1760s, with the invention of a new bottle stopper near Barcelona, did cork get widely used to close wine bottles. That use, and cork's strong seal, quickly became popular. In Paris Thomas Jefferson encountered cork and admired that quality and cork's plasticity so much that for decades he shipped cork acorns to Virginia and tried in vain to grow the trees at Monticello. But cork's finicky nature resisted the transatlantic crossing—the acorns lost their viability faster than the ships could reach the far shore.

* * *

The civil war in Spain was shaping up as a trial run for the ideological contest roiling Europe, between fascism and communism. Germany's new National Socialist government was sending support to Franco; the Soviet Union was arming the socialist Republican side. Crown Cork and Seal's factory in San Feliu de Guixols was in Republican territory, where unions seized factories from the owners. Crown's other plant in Spain was south of Seville, in the region held by Franco.

The war's human cost had become devastating. Yet amid the divisions and destruction, Marsa managed to maintain production, move harvests for export to Baltimore, and make the whole operation work. In his first year he turned over roughly $300,000 in revenue to Crown Cork's headquarters.

Armstrong Cork, which also had a large footprint in Spain's Republican territory, saw its operations there go into limbo. Many foreign companies with factories in Spain wondered what would happen if the Republicans won the war, with communist support and influence. Would all foreign businesses be nationalized? Were any US company's assets safe?

Germany had one of the world's largest cork industries, behind America's, estimated at more than 3 million RM ($1.15 million, or about $20.6 million in 2016 dollars), and German companies had a large footprint in Spain too. In Barcelona, the Nazi Party's representative had begun in 1934 spreading propaganda about a new European order. He worked from an office that also served as a trading company, from which he sent coded messages to Berlin. In two years Nazi groups had spread across the peninsula and quadrupled in number. One of the highest-ranking Nazi officials with interests in the cork industry was Joachim von Ribbentrop, the diplomat who became Hitler's foreign minister. For von Ribbentrop, investing in the cork industry served his holdings in his father-in-law's champagne business while supplying Germany's military needs. Other Germans with cork holdings in Spain and Portugal included the Greiner family of Nurtingen; Kurt Walters, who worked with Armstrong Cork; and Werner Michel Schultz, who worked for years in nearby Palafrugell and whom the Allies later suspected of being an SS officer. Schultz worked under the cover of the cork business and later demonstrated his knowledge of the industry by naming McManus and Armstrong among the world's largest cork businesses. The Crown Cork plant in San Feliu had been German-owned until the 1920s, when McManus had bought it.

In that first year of the Spanish war, Seville's sleepy parochial facade masked currents of change, but one day Gloria saw four policemen on the corner instead of the usual two. She noticed the war's impact in reports of rising crime. Her father talked about cork shipments being set ablaze at the border, and he regarded Franco as a stabilizing force.

After they settled in, they received visitors, including Charles McManus and his son, who stopped by on their way from Morocco and Algiers, before heading to Portugal. With the war engulfing the north and east of Spain, Seville was probably as far north as the McManuses went.

In February 1938, newspapers reported that German planes were bombing the northern Catalan coast, continuing the campaign that had destroyed Guernica months before. The quiet forests of northeastern Spain shook with early morning explosions. Two cruisers sent bombs raining down on San Feliu de Guixols and nearby Palamos. The shelling set the port on fire, killed a policeman, and wounded three others. American newspapers noted that "these towns are centers of the cork industry," where American firms employed hundreds of men and women. How confusing it was for Gloria to be an American, with her father an American businessman and her family's roots deep in the Catalan region being pounded by violence. She knew there was not a point in her where the Catalan stopped and the American started.

* * *

One day when Gloria got home from the French School, her parents were waiting for her. They would be taking her on a trip to Algiers, they explained, where her father had business. They would all go together and make it an African adventure. They would be in Algiers for a couple of months perhaps.

The trip involved a short sea crossing. As Algiers appeared above them suddenly on the coast, the city's white walls shone like a lighthouse in the sunlight. Their car wove uphill from the harbor, and Gloria's excitement rose. A white plaster city perched above the blue sea, Algiers had a very different energy than Seville—it was a modern and bustling place.

Gloria tagged along with the housemaid on the woman's day off each week. Sometimes they visited the market, where vendors huddled with their wares on the steps of the casbah quarter and women passed completely veiled except for an occasional glimpse of patent leather shoes. Other days Gloria followed the maid through narrow streets to another neighborhood in the hills. On these walks vistas opened suddenly through alleys, allowing Gloria to see down into the wealthier homes. She studied the elaborate gardens surrounding the lovely houses. She peeked between two stucco homes to see another palazzo.

Gloria was amazed by the social life in Algiers. She saw Arab men walking with French women—so different from Spain, or even Brooklyn for that matter. To see the races mixing this way was eye-opening.

One evening when her father was going to meet a colleague downtown, he invited Gloria to come along. They would have supper at the hotel where the businessman was staying. She happily joined him, along with her mother and her sister, Mercedes. The drive to the heart of the city was like touring a festival near sunset. The hotel dinner was delightful and sophisticated. But afterward she was startled to find the downtown dark at ten o'clock, as if everyone across the city went to bed at nine.

Other days Gloria was bored, forced to stay inside and read. Algiers was hot, and her family was feeling it. One day, sick with fever, her father rode off to see the cork forests. This puzzled her. Why couldn't he wait until a day when he was feeling better? He drove far out into the arid countryside simply to watch trees growing. In Algeria the cork forests were wilder, often frequented only by herders grazing their animals. Like cork forests all around the Mediterranean, the soil was parched and covered in wisps of brownish grass. The trees offered scant shade.

"Why did you go?" Gloria asked her father afterward.

"It was something I felt I had to do," he replied. Later, when she saw a photo of him taken that day at the edge of the forest, it frightened her. He was seated at a picnic, but he looked deathly ill. She told him so. "With good reason," he said. "I was very sick." Why would he endanger his health that way? It made no sense.

Labor disputes in Algiers were toxic, and the city seemed on the verge of violence. The warnings restricted Gloria's wanderings. It felt unfair to her. The place that had been so fantastical and welcoming had a darker side. The standoff with workers got worse, according to her father, and their business would probably suffer.

* * *

The cork forests in North Africa that had recently seemed promising as an expansion for Crown Cork were already shadowed by Europe's war clouds and local divisions. British intelligence surveilled the forests from Algeria to Morocco, noting the limited resumption of cork shipments between Casablanca and New York. Their strategic importance forced intelligence agents to assess even the old, decaying port facilities of tiny Tabarka, a day's drive east of Algiers at the Tunisian border, and to describe at length landing points there, including two old Turkish forts left from the Ottoman Empire. American spymasters would arrive before long.

Melchor told Pilar, "Maybe it would be safer for you to go back to Spain." If the labor situation in Algiers resolved quickly, he said, maybe they could come back for the last part of the vacation.

Soon after Pilar got the children back to Spain, Seville erupted. Bombs exploded across the river from their home near Torre del Oro. Gloria saw the riverbank smoking from the bombardment and realized it was only by chance that they were on the safe side of the river. From Algiers, her father frantically pulled every string he could to arrange his family's passage to safety. For five days he lived by the telephone. Too distressed to eat, he lived on coffee and seven packs of cigarettes a day.

Pilar and the children were evacuated from Seville, with seventy-nine other foreign nationals, aboard the British destroyer HMS *Shamrock*. When Gloria's father received the cable that they had arrived safely in Gibraltar, he collapsed. Pilar used just four words to let him know they were safe; their Spanish pesetas were worthless, and she could hardly afford the cable. Melchor fired back a reply of four pages, beginning, "The happiest day of my life."

He reunited with them in Gibraltar. Together they boarded a ship for New York. Marsa planned to return after Spain's war was over. But conflict in Europe metastasized.

The industry's channels for exporting supplies thinned to a trickle. Portugal was the only country with a path to ship cork to America, and McManus needed someone in Lisbon with experience to navigate a business climate that had become more treacherous than ever. He refused to accept Marsa's resignation.

The Marsas had settled back into their house on East Nineteenth Street, two miles south of Prospect Park. For Gloria, Brooklyn life felt less bright after seeing the world, and Americans struck her as less sophisticated. Middle school was something of a letdown.

Her father managed his employees in Spain and Portugal from an ocean away, often frustrated by communication delays. He visited the piers where the shipments arrived on the East River and often met Herman Ginsburg, the head of Crown Cork's international operation, for lunch in Manhattan.

Ginsburg too was self-made. A trim and fastidious dresser, Ginsburg stood five feet five inches and had a boyish, intelligent face. With glittering eyes and curly hair swept back, he looked younger than his forty-two years. He had arrived in America in 1905 as a five-year-old from Lithuania but had left all traces of those beginnings behind. The family settled in Ohio, where his father had faced bitter anti-Semitism and found only menial work. When Herman was a teenager, he left high school and the Midwest and made his way to Baltimore. With night school and hustle, he worked his way up to manage Crown Cork's office in Jersey City. Still in his twenties, he was helping run the company's constellation of factories abroad, in close partnership with Marsa.

They made an odd duo: Ginsburg, who was inclined toward the Spanish Republican cause, and Marsa, who voiced respect for Franco. Both had come to America in their youth—within a year of each other. Each had made his own path as an outsider in American business. Each knew that if he had tried to immigrate twenty years later, he might have been blocked from entering the country. In 1924 a new immigration law slashed the numbers of new arrivals, a policy targeted to block people from Southern Europe (like Marsa) and Eastern European Jews (like Ginsburg). Marsa had kept his Spanish passport for decades, but in 1924 he became an American citizen.

In New York the two men became closer than coworkers, often getting together for dinner with their wives. In a photo taken around 1939, Marsa and a fresh-faced Ginsburg are with Ginsburg's wife, Bobbie, at a restaurant. The three are at a table, smiling at the photographer (presumably Pilar), amid the hum of other restaurant patrons. The viewer sees two generations of businessmen, released from their pasts, confident in their future.

After Marsa had been back from Spain for some months, McManus cajoled him to return to Europe for another assignment, this time in Portugal. Again, Marsa considered Europe too unstable for the family. He planned to go solo. Again Pilar refused to let him. The family would go to Lisbon together, whatever happened.

While Marsa and Ginsburg adapted Crown Cork's operations to the European conflict, Charles McManus was positioning the business toward the US government's defense response. As the American economy emerged from the Depression, the government discussed new equipment needs with aviation giant Glenn Martin and other business leaders. Families living near Crown Cork's factories would see new opportunities and pressures.

## Chapter Three

# THE DiCARAS IN A BIND

## · DiCara: 1939–1942 ·

> I want people to read and know and get the feeling of what we *felt*. . . . I want people to know what we went through.
>
> FRANK DICARA, November 2013

Frank DiCara was racing down the alley behind his family's house on Pratt Street, past the outhouses and backyards. Sometimes he ran as part of a game. This time he was escaping. He and his friend Tony had been admiring the fig tree in the backyard of a neighbor, Anthony Castagnera. The old atheist took care of the tree, and it had produced a nice bunch of figs—so nice that Frank and Tony admired them off the branches where they were ripening and right into their mouths. Their timing was bad—Castagnera came out and started chasing them down the alley.

Frank had to use all the obstacles that a Highlandtown alley presented—pigpens, fences, outhouses, and clotheslines—to get away. He was a master.

Thirteen-year-old Frank rarely got to experience the pop of a cap coming off a soda bottle. Like many families in Highlandtown, the DiCaras didn't have spare change for buying soft drinks. He was the youngest of six children crammed into the little house on Pratt Street, with no heat, no phone, no indoor toilet.

Frank's father and mother had come to Baltimore on the same ship from Sicily more than thirty years before. At Ellis Island they had answered questions about how much money they had brought (little), whether they were visiting a relative (yes), whether they were polygamists or anarchists (no to both). Both brought pieces of the Old World with them. Thirty years later, Giuseppe went by "Joe" and made a living as a track laborer on the B&O Railroad, but he kept a vegetable plot in the community garden in Bayview and made wine in his basement. Rosa managed the household and got the children to school. The four boys shared a bedroom, and Frank's sisters shared another. Their parents had the third room.

Joe DiCara had a white handlebar moustache and played trumpet in the band of Our Lady of Pompei Church. He was a community stalwart and a strict Sicilian father. He would tell Frank, "You be home by seven thirty, and sitting on this curb." Every night he would check on the kids in bed and say, "*Bacio la mano*" (I kiss your hand). Their luxuries consisted of a decade-old wind-up Victrola and a Philco radio with a shortwave band on which Joe listened to the heavyweight fights: Primo Canero, Tony Galento, Billy Conn, and Max Schmeling.

When Joe's temper flared, the house felt like a battlefield, and the dinner table could be a piston chamber. The kids knew the look on his face that meant, "Go to the corner and don't move." Frank saw the exhaustion in his father's face when he came in from work, and the boy knew not to wish out loud for snow. When it snowed, the railroad crew came at 3 a.m. to rouse Joe to clear the tracks. He was paid two dollars a day.

The boys chipped in to help make ends meet, with the oldest, Angelo, contributing his pay from a job on Sparrows Point, and young Joe adding in his income from selling fruit and sundries in the Baltimore tradition of horse-drawn vending carts. Frank crammed in odd jobs before and after school: shining shoes, working in a bowling alley, delivering papers across town.

A Highlandtown alley was a world away from the rest of Baltimore. At night it felt darker than any place on earth. Things happened back there. One night Angelo came out of the alley in a state of excitement: he had seen the blessed Virgin Mary on the wall, as plain as day. In their neighborhood, miracles and cures from the Old Country existed alongside the mysteries of American life.

Pratt Street held its share of those mysteries. Next door Miss Trotta ran a bar called Trotta's, where customers arrived quietly but the music could go late into the night. *Fairies*, *queers*, and *lesbians*—those were the only words Frank had for describing the patrons. But he greeted them when their cabs pulled up, opened the door for them, and they tipped him ten cents. Frank heard them playing songs like "Pistol-Packin' Mama":

> Drinkin' beer in a cabaret
> Was I havin' fun
> Until one night she caught me right
> And now I'm on the run
> Lay that pistol down, babe
> Lay that pistol down
> Pistol packin' mama
> Lay that pistol down

In warm weather with the windows open, the music kept him awake until two in the morning. Miss Trotta also let Frank come and clean up the next day, when the empty bottles and floor smelled like stale beer and worse. The boy could keep any nickels and dimes he found. "They're all yours," she'd say. "Just sweep it all up."

Highlandtown didn't rate a mention in most guidebooks to Baltimore. The only one that mentioned the neighborhood was the WPA guide to Maryland. The local authors of that guide, comparing the DiCaras' neighborhood with Little Italy, called Highlandtown "larger, though not as picturesque." (The guide also said that Baltimore's character grew from the "lusty, cantankerous life" that roiled "under the surface of its neo-British aristocracy.") But Highlandtown had traditions. During the festival of Our Lady of Pompei, men hoisted a statue of the Virgin through the streets before crowds of devotees. Along the procession route, as the band played, people pinned money to the figure—five- and ten-dollar bills—or to the pillow holding it, seeking favor. Men hoisted the Madonna from the sanctuary and carried her through the streets, starting from Claremont, north on Conkling, and over six blocks to Pratt Street, past the DiCaras' home. The brass band moved slowly and noisily closer, with Frank's father's trumpet sounding loud. The music trailed through the blocks that formed the world that Frank knew, from his cousins' homes in the row houses off Eastern Avenue to the south, to Grundy Street to the east, and up Lombard to the north. Frank had friends west to Patterson Park, where you could set out a folding chair and sleep through a hot summer night under the trees without anyone bothering you.

Over the years, Joe DiCara stepped aside from the band, and his son

Angelo took his place. When the Pompei band came around to Pratt Street, with Angelo on the trumpet, their father would take his glasses off and wipe his eyes, moved by time's passage. Joe DiCara was a tough old Italian, Frank said, but he kept things in his heart.

* * *

When Frank went to bed on the night of September 17, 1940, Baltimore's economy was on the verge of recovery, and Highlandtown had lots to occupy a thirteen-year-old. On Saturdays he and his friends would go to the Rivoli movie theater uptown, where a ticket, a Pepsi, and a hot dog cost six cents. The theater's decor had the gaudy feel of the burlesque shows that occupied that block of Baltimore's tenderloin. At the Rivoli, Frank would soak in the latest thriller. In *The Human Monster*, Bela Lugosi played a London insurance agent who murders his clients, using a home for the blind to cover his scheme. Before the feature, newsreels showed the latest horrors from Europe. They blended almost seamlessly into the film's inky black opening images of London Bridge, with bodies bobbing on the tide.

After the film ended, Frank stayed and watched it again.

That night from the boys' bedroom, Frank watched the glow in the distance with its strange pillar of smoke.

"I looked out my window and could see it burning," Frank told friends later. "It burned for days."

The pillar of smoke from the Crown Cork and Seal factory hung over neighborhood conversations, threatening more than property. Everybody felt the war jitters from Europe. Frank heard the rumors of sabotage. "Somebody started the fire," he'd say. "It didn't start by itself."

At the dinner table, Frank's father said the fire had started on a railway siding.

That fall, like every year, Joe DiCara made wine. At the start of the season, Frank sat in the window and watched his old man out in the yard take a candle and char the inside of the Weiskettel's barrels. A truck pulled up, and his boys formed a line to unload fifty-plus boxes of grapes to the basement, where Joe kept a torque press. He squeezed out three barrelfuls a season—what he called first wine and second wine. Frank wondered at the care and expertise that his father showed in winemaking. How was this man in his sixties still working as a laborer on the railroad?

When cousins came for dinner, Joe would send Frank to the basement to fetch a bottle. Family was *always* around, playing music and talking. Frank's father would deal hands of *Sett' e Mezz'*—seven and a half cards. One uncle would bring his accordion and play late into the night.

Highlandtown sometimes felt separate from the rest of the world, like when Frank read in the comics a strange new catechism from *Blondie*:

> E's for the Enemy
> Who's our deadly foe
> Jap, Nazi, Italian
> Are three in a row.

The newspaper never printed anything about Highlandtown except in the crime pages or when a judge denied citizenship to applicants from Italy and Germany. Or when ships from Italy, anchored in Baltimore Harbor, were sabotaged by their crews, who didn't want to return home to Mussolini.

On winter mornings Rosa DiCara woke at 5 a.m., took a hand ax, and chopped wood to start the potbellied stove. When the kitchen was warm and it was time to wake the kids for school, she grabbed a broomstick and headed to the boys' room. The house was so cold that Rosa had to use the broomstick to get Frank out from under the covers. He dressed, stuffing cardboard in the holes of his shoes to make them last until the next time his mother could afford a trip to Epstein's and haggle down the price. *Pair of shoes fifteen cents? No, it's seven cents!* She worried every dime, fed the children, and at the end of the day she often went to bed hungry.

One of the traditions his parents brought from the Old World was a remedy for Rosa's severe headaches. She said they were caused by *mal occhio*, or the evil eye, which seemed to be getting worse. Joe would take a dish of water and put some olive oil in a smaller dish, then dip his thumb in the oil to form drops in the water, until he made seven of them. At seven drops, the old wisdom said, the evil eye and the headache would go away.

Rosa stretched Joe's laborer's wage as far as she could, feeding the kids calamari when a man with a scale came around selling it for five cents a pound, along with minnows that she would fry up like chips.

Some days she sent Frank through the alley to Grappalo's to buy lunch meat on credit. Mrs. Grappalo marked the amount in a brown book for payment due at the end of the week.

Everyone needed to do more. So Frank walked six blocks over to Marks' bowling alley on Fleet Street and applied for a job. The owner, Jimmy Marks, hired him to set up the pins, a crazy introduction to the work world. Frank's task was to wait at the end of the lanes for bowling pins to come crash around him and then to reset the pins until they crashed again. It was explosive, with the hardwood clubs flying everywhere. Every minute, Frank was going back to the treadle (or rack) to set the pins again. He stepped on the treadle to make it go up and wrestled the pins into position. What really sent fear up his spine were the speed bowlers. They sent the pins flying with blistering and painful speed.

Every week Jimmy Marks paid Frank in pennies wrapped in brown envelopes. Frank sealed the flap, walked home, and gave the envelope to his mother unopened.

Highlandtown could be a frightening place, but Frank was also learning that some things weren't so scary, like gambling. Elsewhere in Baltimore, gambling ranged from tavern pinball to the Preakness and casinos that could handle a $1,000 note. But in Highlandtown, the options narrowed to street-corner crap games and bookmakers. Benjamin "Benny Trotta" Magliano was a boxer who ran a bookie operation. Frank knew that Magliano and his family kept a radio inside the piano, which they used to follow race results. Frank got pulled into helping Magliano's bookie operation by keeping lookout. If he saw a policeman, he was supposed to run and yell, "Six to five!" to let them know the cops were coming. When police showed up unannounced, Magliano's crew quickly popped the betting sheets in their mouths.

Rising wartime fears raised the barriers between Highlandtown families and the rest of the city even higher, limiting their options for making a living. In a courtroom downtown, judge William Coleman denied nearly three dozen Italian and German applicants for citizenship. He pointed to Europe for justification. The applicants had expected a naturalization ceremony, but he shocked them, saying, "I must say to persons still owing allegiance to Germany and Italy . . . naturalization will be postponed as to you until further notice."

Tearful applicants left the courtroom in despair. Many were haunted by the trial of factory employee Michael Etzel, a twenty-two-year-old

Baltimore man who had worked in a plant making bomber planes. Etzel was tried for sabotaging two dozen B-26 Marauder bombers on his assembly line and was convicted of using a fishing knife to cut wires and hack rubber tubing to the planes' gas tanks. His motives were muddled. At the time of his arrest, he said he did it to protect relatives in Germany. On the witness stand, he insisted that his motive was to "spite the Glenn L. Martin Company" for treating him unfairly. At Etzel's sentencing on November 18, 1941, the judge called his crime "tantamount to attempted murder" and ordered him to serve fifteen years in prison, an unexpectedly harsh sentence. It was intended as a warning to others.

Etzel and his wife lived less than a mile from the DiCaras. The neighborhood was in the crosshairs of federal scrutiny. The government was still considering whether to detain and relocate Italian Americans and German Americans, as it did with Japanese Americans. One of the main immigration detention centers was Fort Meade, just southwest of Baltimore.

At age thirteen, Frank didn't think too much about news beyond the headlines of the newspapers he delivered. He forgot his suspicions after the Highlandtown factory fire and was busy with his after-school jobs and classes at Our Lady of Pompei, where the nuns were tough. When he could, Frank and other kids from the block would hang out in front of a confectionery store on the corner. One girl in that group named Irma lived around the corner. She would come down to the stoop, and if Frank was there, she'd stay.

* * *

After Pearl Harbor a new US intelligence agency emerged, the Office of Strategic Services, to fill the gap that had made such a surprise attack possible. The OSS had responsibilities for sleuthing overseas but also at home. Sometimes the OSS worked in parallel with the FBI, with a branch reserved for "foreign nationalities." Other times the two agencies did not coordinate.

OSS agents visited Highlandtown to sound out Italian American residents on their sympathies for America's enemies. At the same time, the FBI mapped Italian communities across the country, targeting ninety-eight "Italian aliens" for arrest within days after Pearl Harbor. The largest concentrations of those arrests happened in New York, California, Virginia, and Baltimore, where ten people were taken into custody. The OSS agent in Highlandtown canvassed Frank's neighborhood, chatting up residents and listening for fascist sympathies. After his tour, Agent L. R. Taylor slipped a page into his typewriter and tapped out: "My inquiries were made in grocery and fruit and drug stores and from some people I met on the street. A man and a woman down at the dock talked to me at some length. In several shops I was entirely unsuccessful in eliciting any opinion. The answer I received in one case was something like this: 'I am a good American. I have been here thirty-seven years and I don't know what's going on in Italy and I don't care.'" The Highlandtown residents Taylor spoke to knew with whom they could safely talk. Everyone had relatives back in the Old Country and was concerned for them, but the times had become treacherous. If pressed, some like DiCara's father voiced a wish that America and Europe would simply mind their own business.

Before the OSS began its domestic monitoring, intelligence agents from the European Allies had tracked foreign influences inside America in a very different direction—among American reactionary nativist groups like the America First Committee. Before Pearl Harbor, British spies were convinced that if they could prove that America First and other isolationist groups were receiving Nazi funding, the evidence would unsettle the American public enough to push the United States into the war. British spies found those ties: Germany used key contacts in New York, Chicago, Washington, and San Francisco. And in Cleveland and Boston, they traced money transfers from Nazi sources to America First.

Before the British could issue these findings, though, Pearl Harbor was attacked. America was in the war.

* * *

Frank saw changes after Pearl Harbor, mostly outside his neighborhood. Overnight, port security across the country tightened. The Coast Guard recruited volunteer patrols and put them through night classes on fire prevention, the ins and outs of ships and docks, and countersabotage. The volunteers, many of them women, received marksmanship training and .38 pistols.

Silence and secrecy were the new watchwords. Posters went up: above a drawing of smoking wreckage at sea hung the words "Loose Lips Sink Ships." Other placards warned "Someone Talked!" Snapping photographs in a port or airport could get a person arrested.

Behavior suddenly looked different. Actions that were innocent before now received glares of suspicion. A twenty-year-old Baltimore shipyard welder worked faster to boost his paycheck. The FBI charged him with sabotage, accusing him of intentionally doing shoddy welding.

Streetcars sailed down Eastern Avenue with posters on their sides urging people to enlist. The 26 streetcar rattled in from Sparrows Point, depositing each shift's crew from Bethlehem Steel, and continued on to Bay Shore, where on weekends Frank and his friends went to go swimming. At Sparrows Point Frank's brother had worked his way up to foreman. Half of Highlandtown worked on the point, where Bethlehem Steel employed almost fifty thousand people. Most were white; about 15 percent were African American.

The dynamics in the Sparrows Point factories were shifting. For decades management had been biased against black workers and suspicious of unions. In 1941 unions organized to get Bethlehem Steel workers to vote on unionizing, and the discrimination faced by black steelworkers was a factor. Subject to the vagaries of a white supervisor, a black worker often didn't get overtime pay or breaks. When the vote at Bethlehem Steel was held in May, steelworkers won, for the first time, the right to have a union at Sparrows Point. Nick Fontecchio, the union's district director, told the *Afro-American*, Baltimore's black newspaper, that black workers played a decisive role in the vote: "We appreciate their support and we're going to work for the common good of all," he said. The relations between workers and management remained tense despite a united stance to meet wartime production targets.

America faced war on two fronts, and that meant more jobs in Baltimore factories. President Roosevelt set impossibly high targets for American industries: 60,000 aircraft needed to be built in 1942 and 125,000 in 1943. Baltimore bore a large part of that burden, along with Detroit, producing planes like the B-26 Marauder, one of Martin's most successful bombers.

"Powerful enemies must be out-fought and out-produced," the

president told Congress weeks after Pearl Harbor. "We must out-produce them overwhelmingly, so that there can be no question of our ability to provide a crushing superiority of equipment in any theatre of the world war."

The city's other industries were scaling up too, and people were coming long distances for jobs, not just at Sparrows Point but also at Glenn L. Martin, and at Crown Cork and Seal. It could be hard to know where to apply, since new factories were springing up and they were often camouflaged. War factories went to extremes to outfox the enemy. The Bendix building was camouflaged with a trompe l'oeil paint job.

With so many new people coming to the city, strange happenings caused whispers of sabotage. The Maryland Sanitary Manufacturing Company, just east of Highlandtown, burst into flames in the middle of the night. The next day nothing was left. All that people knew was that the company was working on defense contracts.

* * *

While things changed after Pearl Harbor, the bigger dividing line for Frank's life was when his father got sick months later in 1942. During the summer, like other years, Frank helped his father prepare to make homegrown wine. Now with his brothers all in the service, it was left to Frank to haul in the crates of grapes. He carried them to the basement and pried them open. Frank weeded the garden in Bayview, where his father grew vegetables. Life had routines: the morning walk to school at Our Lady of Pompei, after-school jobs, and street games. Frank played Paddlestick, where you hid a paddle in the alley and whoever found the stick would use it to beat the others. For Hunchback, four kids lined up and others jumped them, sort of like leapfrog, except instead of making a chain, the idea was to see if the line could hold your cannonball weight or if it would collapse.

Frank was spending more time with Irma and the group at the corner confectionery store. She lived on Mount Pleasant Avenue, and her father, it turned out, was Anthony Castagnera, the old atheist with the fig tree. Irma's father came from Sicily, like Frank's parents, worked as a bricklayer, and had gotten his citizenship papers a year before. (Frank's father had put in his citizenship papers, but his mother never did.) Irma went to Patterson High, not Our Lady of Pompei, and her family didn't go to church at all. How could that be? On evenings when Frank met Irma on the corner stoop, they talked until her father called out, "Irma, come on home."

At home, Frank got into trouble and faced friction with his father. One time he and Tony found a stash of moonshine that Frank's father kept, bought from a guy who came around selling it from a Model T. The boys started playing cards for drinks, and Frank was winning. Winning so much that he passed out. His sister found him in the kitchen and dragged him upstairs, where he threw up. Another time the boys, for lack of cigarettes, wrapped corn silk from cornhusks and smoked it. They puffed and blew out the smoke.

That's when Frank's father caught them. Terrified, Frank bolted to the backyard, where his father chased him and grabbed a clothesline pole. He raced after Frank and back up the alley. Running past a gate, Frank swung it open and his father hit it, knocking off his glasses. Pacing the block across the street from their house, Frank caught his breath and saw the old man in the living room, waiting for him. Hours later Frank got up his nerve, crept around the back of the house, and scaled a drainpipe up to a second-floor window, his sisters' room. They let him in and he crawled into bed between them. Time went by. It got dark. When their father made his night rounds and checked on the girls, he saw Frank in their room. He said, "*En gadda garda yo. Domani mattina.*" Meaning: "Watch yourself. Tomorrow morning, you'll hear about this." Frank couldn't sleep that night.

From time to time, Frank passed the Crown Cork and Seal factory, which was practically a city unto itself. Sometimes he stopped and watched from the Eastern Avenue Bridge as the workers went inside and the big green trucks came out of the gate. From the bridge he could almost see the cranes in the port where the cork freighters docked. All the mystery of the real world beyond Highlandtown.

* * *

Weeks after Pearl Harbor, families from German and Italian backgrounds, in Highlandtown and across the country, learned that the government had branded them "enemy aliens." Now they were required to register at the post office and to carry pink booklets labeled "enemy alien." The government program managing that registration process, out of Philadelphia, was the same agency that moved Japanese Americans into work camps. Baltimore swirled with rumors that the government was rounding up families and detaining people at Fort Meade. Right after the Japanese attacked, federal agents across the country arrested ninety-eight Italian "aliens," including ten in Baltimore. The agents identified their targets with the help of the Census Bureau. They didn't need warrants. Having sons in the army, or serving in the military yourself, was no guarantee of safety. Across the country, some ten thousand Italian Americans were detained or relocated. One navy serviceman came home on leave to find his family's home vacant and boarded up, his parents gone. The situation was too tense for adults to discuss in front of children.

Two months after Pearl Harbor, FBI agents arrested 264 Italians, along with nearly 1,400 Germans and more than 2,200 Japanese on the East and West Coasts. That February President Roosevelt signed Executive Order 9066, allowing Secretary of War Stimson to arrest and imprison Germans, Italians, and Japanese declared enemy aliens. They could be held without charges or trial, and their homes and businesses could be summarily seized.

Amid these tensions, Highlandtown came together for events like I Am an American Day on the third Sunday in May. William Randolph Hearst had championed the day for years as a national holiday to celebrate citizenship. At the World's Fair in New York, an ad man had promoted a song titled "I Am an American," and Congress made the holiday official in 1940 as a response to rising concerns about illegal immigrants. The holiday became a way to recognize naturalized Americans and the importance of citizenship. Warner Brothers made a propaganda film, *I Am an American*, promoted by the Immigration and Naturalization Service. On that day in May, Highlandtown veterans and politicians paraded through the streets, and bystanders craned their necks to see.

But parades didn't stave off the suspicions of people outside the neighborhood. Italian American families still felt they had to prove their loyalty. As military recruitment rose, thousands of young men from these families enlisted. In the war's first months, seventy-five thousand military inductees were Italian American. (Throughout the war, that figure rose to between a half million and one million.) One Sicilian-born attorney who chaired his local draft advisory board boasted that more than half the young men drafted from his district were Italian American—well beyond their proportion of the general population.

Still the suspicions intensified. One morning in the spring of 1942, federal officers knocked on the door of a home in New Haven, Connecticut. The man who opened the door, Pasquale DeCicco, was a pillar of his community and had been a US citizen for more than thirty years. He was taken to a federal detention center in Boston, where he was fingerprinted, photographed, and held for three months. Then he was sent to another detention facility on Ellis Island. With no hearing scheduled, he was moved again to Fort Meade. On July 31 he was formally declared an enemy alien of the United States.

He had done nothing wrong, DeCicco argued in a letter to the attorney general. "My concern had always been the maintenance of cordial relations and friendship between the United States and Italy," he said of his work helping Italian immigrants to navigate the US legal system. He only wanted the two countries to understand the other's "habits, traditions . . . and freedom, their lofty ideals."

DeCicco's appeal failed. He remained at Fort Meade until December 1943. He was never shown evidence against him nor charged with any crime.

In Illinois, a young postgraduate sociologist named Paul Campisi documented the growing unease in his Italian community. Campisi saw a tremendous "fear, bewilderment, confusion and anxiety," he wrote. There, too, rumors began soon after Pearl Harbor: the government was going to pass a law taking away the property of all Italians who didn't have citizenship papers; Italians living near defense factories would be forced to move; Italian homes would be searched, and cameras, shortwave radios, and guns would be confiscated. In fact, officials were considering all three of those options.

Campisi found a contrast between how the older Italian-born generation saw the threat and how second-generation Italian Americans viewed it. The older generation felt a deep inner conflict. "It was hard for the Italians to believe that their homeland was actually at war with America. It was incredible, unbelievable," Campisi wrote. They had had to register as aliens following the 1940 Alien Registration Act, a process that had filled them with anxiety. Yet nobody believed it would go any further. Nobody expected the jolt of December 8. Campisi wrote, "It was a dual reaction. First, anger, amazement, and incredible shock at the news of Pearl Harbor, and then sorrow and pain at the realization that Italy definitely would now be an enemy nation." Now they faced even greater suspicion from their coworkers and friends; even worse, "there was great sadness that Italy had to be an enemy nation, and that henceforth all things Italian should be suspect and hateful."

"I never expect to go back," said one Italian who had lived in America for nearly forty years, "but I can't help but feel sorry for relatives and affectionately of my home village."

"*Povero* America," Frank's father said. "Poor America, you ought to stay home and take care of your own house." Europe wants to be communist or fascist? That's up to them. Why should we try to make them any different? Politics came up at the dinner table more often.

That spring the DiCaras must have gotten a notice in the mail. They couldn't put it off any longer. Everyone was getting them. The whole situation seemed idiotic, and Frank resented having to go to the post office and register. His mother paged through the papers again. Everyone over fourteen years old had to answer forty-two questions, concerning everything from when they first entered the United States to their employment history and religious affiliations. Walking to the post office, Frank resented everyone—the people on Eastern Avenue coming out of Epstein's looked carefree and untouched by this ridiculous bureaucracy. Others seemed to smirk, laughing like they knew the humiliation of his situation. By the time he reached the post office on Calvert Street, the big gray stone building looked like it would crush them. How the hell could his mother let the government stamp a card calling her an enemy alien? She and Frank's father were hardworking citizens, decent parish members at Our Lady. They had sons in the US Army. And his father's health was getting worse. Frank felt like he was in a horror film. So he shut it out of his mind.

* * *

Frank churned through the days in classrooms on an upper floor at Our Lady of Pompei. Graduation couldn't come soon enough. He knew his family couldn't afford to send him to college, and he didn't know anybody who had gone. There weren't scholarships for kids like him. But he could make his way up in the world with a factory job, even though those jobs were getting harder to come by.

Highlandtown families, like immigrant families elsewhere, were seeing the effects of their enemy alien status. Some cities began to drop foreign-born residents from the rolls of county relief agencies and to deny them charity. Many businesses fired workers who weren't naturalized citizens. Fort Meade, the sleepy military-base-turned-immigration-facility, began to house more detainees.

Then on September 22, 1942, a short item appeared among the death notices in the *Baltimore Sun*: "DiCARA—On September 20, 1942 GIUSEPPE, beloved husband of Rosa DiCara (nee Cavallaro). Funeral from his late residence, 3800 East Pratt Street, on Wednesday at 8:30 A.M. Requiem High Mass at Our Lady of Pompei Church at 9 A.M. Interment in Holy Redeemer Cemetery."

Frank's father had a spartan funeral at the house. In the decades since he had arrived on the *Calabria* from Palermo with less than $50 in his pocket, unable to read or write, he and Rosa had made a life in America. After the gathering on Pratt Street, there was a Mass at Our Lady of Pompei. Joe DiCara was buried without ceremony at Holy Redeemer Cemetery.

Frank's brother Angelo could not get his leave request approved by the army to attend the funeral. So friends in his unit covered for him and he went AWOL just long enough to see his father buried.

Frank was confused to see Angelo home, and numb. The drive out Moravia Road to Holy Redeemer Cemetery took twenty minutes, but it felt like a million miles. The car bearing the casket rode through the iron gates and up the gentle slope to where the newest graves were, amid the fall leaves.

It seemed like his father's passing sealed away the Old World, its seasonal rhythms, family gatherings, and good nights. The circle of family in Highlandtown gave Rosa some support, but Frank felt alone. His brothers were away in the service, and he was supposed to be the man of the house. His sisters, both excellent seamstresses, took the only jobs they could get at a low-paying sewing factory called Isaac's.

Breadwinning pressure landed on Frank, and he felt it in his stomach. He asked for a raise at the Esskay box factory, where he worked after school. But his boss wouldn't budge—no raise. So Frank quit school to find a full-time job. Not that he was a great student, but he had expected to graduate. It struck him as unfair that as jobs opened up and Baltimore had more wartime manufacturing, there were no other DiCara brothers around to take those jobs. Only Frank.

Through a friend of a friend, he got a job at a factory assembling wings for bomber planes. The pay was good. By January Frank was walking a mile to the factory at Boston and O'Donnell Streets, exhaling a cloud of wintry air, feeling like an adult. He was fifteen years old, making sixty cents an hour—four times what he'd been making at the box factory, and full-time.

Irma was studying stenography at Patterson High. It was a way to get ahead. Through her, Frank saw other ways of living. Her grandfather built homes. Frank imagined having a job like that, with a middle-class life. When he took Irma to the movies at the Rivoli, he felt great. The feeling got him through the next week.

Each morning, Frank left home with a sandwich his mother made and walked the ten blocks to the plant. He passed other young guys playing craps on the street corner, some who ran numbers, keeping their eye out for Emerson the Maul. (Sometimes the undercover cops joined the crap games and stole the kitty.) The people he'd grown up with had carved out unexpected niches for themselves, like the Zannino boys, whose family ran the funeral parlor near the church. Everyone in Highlandtown went to Zannino's for funerals. Frank knew that some families had ties to the mafia in Baltimore and that John Zannino had met an untimely death in the summer of 1941. Sam Zannino would disappear ten years later. Even Frank's father had had a dark thread in his past; he had told Frank that for a while he'd been with the Black Hand, an extortion group that faded away in the 1920s.

On Frank's walk to work at the Marauder wing plant, he passed the sweatshop where his sisters worked, making almost nothing per hour. Getting through the security line to the factory was nerve-wracking. After the Etzel trial, the plant had added two security guards at the gate outside and more inside. There was one big guard who Frank passed every day.

Through word of mouth, Frank heard other factory sabotage stories. Everyone knew about how Etzel had cut vital circuits and left a cardboard sign on the navigator's seat in one of the planes. His handwritten sign said in red and blue crayon: "B-26 bomber. Martin's deathtrap. Martin's Government Draft Project. Heil Hitler!" Etzel changed his tune on the witness stand and said he had cut the wires because he was angry at the company for having promoted a junior coworker over him; coming back from a vacation, he had found a greenhorn whom he had trained promoted to be his foreman. "I taught him all he knew, and I wasn't going to have any fellow I had taught bossing me," Etzel said. But he had been stupid and scared people with his testimony ("I had all the opportunity to sabotage planes if I had wanted to"), and now he was in prison for fifteen years.

People whispered about U-boat sightings in the Chesapeake Bay and up the coast. U-boats sank hundreds of merchant ships worldwide in late 1942: 96 in September, 89 in October, and 126 in November. By mid-1943, the Germans had 118 U-boats off the East Coast and in the Caribbean.

The factory was a pressure cooker, with lots of puzzles and mysteries, including who owned it. There was no sign anywhere on the place. When the wings left the plant, Frank knew they went to Glenn L. Martin's factory northeast of town, which employed more than 45,000 workers. But when Frank got paid, his paycheck said Crown Cork and Seal.

* * *

American industry was way behind in achieving the president's production targets. By the fall of 1942, the country was expected to send 5,300 fighter planes to the Allied forces in Europe, but most were not yet built. The need was especially intense for bombers. So the wartime plants started working round the clock. Frank's plant had three shifts, and he followed a rotating schedule: some nights he worked from 11 p.m. to 7 a.m., some days from 7 a.m. to 3 p.m., others from 3 p.m. to nearly midnight.

The assembly plant was huge. It was making some of the biggest items in the industrialized world: vast, shiny wings for B-25s and B-26s.

Frank didn't get much training, but the work didn't require that much. He was made a drill-press operator. He walked the length of each fighter wing—the length of a good-size house—drilling holes for rivets. After he finished one wing, he marched back to the starting point and started on the next. The thirty-foot-plus wings moved along, and he moved with them, popping the drill in at regular intervals along the way. In another part of the plant, they made machine-gun belts.

Just keeping up with the line and keeping out of the way of the union boss kept Frank tense. By the end of his shift, his hands were shaking.

Inside the plant, like at the docks, everything could look like sabotage to the cops. Simply working fast to keep the line moving could bring in the FBI. The agency had not charged anyone with setting the 1940 Crown Cork and Seal factory fire, but in 1942 FBI agents arrested another factory worker at that plant for suspected sabotage of fire extinguishers. Boyd Stalnaker, a fire-prevention worker for Crown Cork, was held for plugging fifteen fire extinguishers and nozzles, and slashing over a dozen fire hoses. "I wasn't getting enough money," Stalnaker said about why he did it.

The same US commissioner who had prosecuted Etzel handled the Stalnaker case. Judge Coleman sentenced Stalnaker to fifteen months in prison. "Whether this man's aim was an act of sabotage or seeking personal advantage, he is a menace to the community," the judge said. It was as if the judge recognized that it could be hard to draw the line between patriotism and personal economic interest.

Baltimore's jobs attracted people from across the South, as well as men who had dodged the draft. At the plant, Frank met people from all over the country, including a woman from North Carolina who was older, in her late thirties he guessed. She wore glasses and had a drawl that Frank hadn't heard before. She grew up a Baptist, she said. That was exotic to Frank. They talked sometimes during the short meal break or over coffee.

One day they took their lunch break together and shared a table in the break room. As other shift workers came and went, they made small talk and she raised the subject of religion. She said, "Frank, you're Catholic. I know what you believe in. I don't believe like you believe. I believe that the end of the world is when you die. That's the end of *your* world."

In her Carolina drawl, these strange words struck Frank with unexpected force. All around them, hand trucks and factory equipment zipped to and fro, and the odd dance of the wings on the assembly line continued just beyond.

What the woman said made him pause. He considered his world, his family and friends and Eastern Avenue. The months since Pearl Harbor and his father's death had shaken them all. His mother cried to herself in bed at night, worried about her sons. All the pieces of the Old World that their family cherished, this new harsh world made jagged and shameful.

"There's not going to be trumpets blowing and all the dead coming back," the woman continued.

Frank thought about that. He said, "That makes sense."

She looked him in the eye. "I want to tell you something. You're a young kid," she said. "Do you want to know what hell is?"

He wasn't sure where she was heading.

"Hell is on earth," she said. "This is your hell. Where can you find anything worse than your father dying of cancer, your brothers going off to war, all of this stuff? Don't you think that's hell?"

Frank considered that too. It was almost a relief to hear someone say it out loud. "Yeah," he admitted, "I have to believe you. That *is* hell."

Frank worked for a millionaire's company, making warplanes flying to Europe, where Frank's brothers were fighting on the ground. Frank saw the odds as stacked against his family. In the first year of America's involvement, historians Robert Zieger and Gilbert Gall have noted, the war pulled millions of working families into tighter bonds with the government and business. The war bureaucracy extended the government's reach into their lives, and the Revenue Act of 1942 increased tenfold the number of Americans owing income tax. The war economy drew workers into defense jobs. The most life-threatening tie, of course, was conscription into military service. These new ties held opportunities and great risk.

* * *

When the president and first lady visited the Ford Motor Company's bomber factory in Willow Run, Michigan, in the fall of 1942, Henry Ford and his son Edsel personally greeted them. Their limousine drove slowly through the factory as Edsel, seated beside them, described the equipment. Security precautions required any factory worker with a lunch pail to put it on the ground out of reach, to avoid risk of concealed weapons. Workers along the route paused and applauded FDR, and Eleanor commented on the large number of women working in the plant.

The visit spotlighted industry's efforts to meet the war production targets, but that September American workers expressed dissatisfaction in a spate of strikes and strike threats. The *Chicago Tribune* declared, "Strikes, Labor Unrest Sweep Country" as steelworkers, truck drivers, miners, and bomber plant employees pressed for higher wages. In Pennsylvania steelworkers walked out, demanding a ten-cent increase in their hourly pay. About 1,200 tire factory workers in Ohio went on strike. Tension between company workers and management was at a peak.

At the end of the Willow Run assembly line, the Air Corps had a row of aircraft from America's arsenal, and the Roosevelts passed the B-26 Marauder, produced in part by Baltimore workers and a bottle-cap maker whose patriotic ambitions reached ever further.

# PART 2

# BUDDING FORESTS AND SPIES

## Chapter Four

# THE McMANUS CORK PROJECT

### · McManus: 1940–1942 ·

> We were three people and a cork oak, but I have to confess that I believed then, and do now, that all in all there were four of us, each one committed.
>
> JOAQUIM VIEIRA NATIVIDADE, Portuguese forester

For the McManuses, the months after Charles Jr.'s escape from Britain in late 1940 were chaotic. Germany continued to sink more merchant ships, throwing trade into turmoil. The Nazi blockade of the Atlantic made all imported products hard to find and much more expensive. Even US businesses that managed to import all the materials they needed still faced the likelihood of wartime restrictions from the government. Many companies lost the slim hold they had on recovery from the Depression. Many went out of business. But for a few industrialists like Charles Sr., things were looking up. Through the Depression he had grown his company to the point where Crown Cork was producing half the world's supply of bottle caps. He had gambled again in 1936 and acquired the Philadelphia-based Acme Can Company, extending the reach of Crown Cork into the more modern tin can business.

Since his return to Baltimore, Charles Jr. was considering carefully his father's blind spots. From the time that Charles Sr. had bought Crown Cork fourteen years before, his idiosyncratic vision literally had shaped the company. Because of his eye problem from the childhood gunshot wound, the elder McManus kept no notes of any kind. His bad eye was an invisible issue in the business.

"If you knew it, you knew," the son explained later. "He didn't talk about it."

Junior had heard his father tell the story about the stray gunshot in school, the many doctor visits afterward, the operations to remove most of the shrapnel, and the painful recovery. His father never let on that the eye still caused him trouble and pain. The old man had a driver's license and drove everywhere, but as soon as his son learned to drive, McManus Sr. said, "You wanna drive? You drive." And he turned the wheel over to the boy.

That was when Charles Jr. realized that his father was having real problems in addition to his usual quirks. He also used a verbal shorthand that only people close to him understood. At the factory, sometimes a staff member would approach Junior with a worried face, asking for a translation of what the old man had said. "Your father said twelve. Did he mean twelve or did he mean twelve dozen?"

"You have to watch Dad when he gets into numbers," the son would say. He liked to tell a story from when he was a boy in the days before Prohibition, when they lived in Manhattan. Back then, his father sometimes brought him along on errands, and one day in January 1920 they stopped in at a liquor store on Broadway near 155th Street that his father frequented. Like every other liquor dealer that month, the shop was ready for customers who were stocking up on alcohol before January 16, when the Eighteenth Amendment went into effect. The dealer was used to a bottling-industry shorthand in which "twelve" meant a gross. That day McManus was talking fast—ordering twelve of this, twelve of that—and the shop owner mistook his meaning and sent twelve cases of champagne instead of twelve bottles. That stock of champagne lasted the family for decades.

At Crown there was no real handbook or procedures in place. Some business decisions his father made were merely inspired hunches. Others were inscrutable. Senior had acquired businesses across the United States that seemed to have little connection to bottling or containers, including a company named Detroit Gasket. (Junior all his life pronounced the word with heavy stress on the first syllable: *De*-troit.) Crown Cork manufactured composition cork, cut it into sheets for automotive uses, and shipped it to Michigan, where Detroit Gasket stamped out designs for the auto industry. Henry Ford was its biggest customer.

Other decisions were sentimental, like the purchase of a factory in Algeria that Junior felt they never should have bought. The plant manufactured wine corks only and shipped them to France and Germany. It had been owned by a cork man in Philadelphia named Larry Johnson. When Johnson died suddenly, McManus felt sorry for his widow and bought the factory. The plant had brought only headaches.

His father also decided to open a factory in southern California. A few months after Charles Jr. and Mary returned from England, thankful to be alive, they went to Los Angeles, where the old man sent him to help set up the new factory. He argued against the move—there was too much to do in Baltimore for him to go off to California—just as he had protested a year before that he wasn't experienced enough to manage the London operation. The old man ignored both arguments.

In the spring of 1941, McManus Sr. arrived in Los Angeles for a mix of work and pleasure. Crown Cork had two plants in California under its subsidiary Western Stopper Company: a primary plant in San Francisco and the new one in LA. The old man wanted to check on Charles and Mary and see how they were settling in. The trip was a welcome distraction from the problems of unraveling supply lines for importing cork. The Germans sank scores of merchant ships every month. According to rumors, U-boats lurked just off Maryland's coast and saboteurs skulked in the harbor, prompting a boost in harbor patrols and factory security. The Etzel trial had also stirred up anxieties in Baltimore.

Moreover, the Selective Service Act of late 1940 required all men between twenty-one and thirty-five to register with their local draft boards. It brought the war closer to the family.

* * *

The military classified Charles Jr. as 4F due to his bad left ear—not surprising but still disappointing to him. He reminded himself that his bad ear had led to the best part of his life, even though his mother never quite approved of Mary's blue-collar family. Instead of serving his country, Junior was blinking in the southern California sunlight while his younger brother, Walter, prepared to be inducted into the air force. Other people made plans as if life were sailing along normally. In two months Walter would be standing in a garden in Maplewood, New Jersey, getting married, and Junior would stand beside him as best man, like George Bailey in *It's a Wonderful Life*. Walter teased his brother: How could he be living next to the world's greatest surfing and sailing spot yet never dip a toe in the water?

Junior and Mary saw how quickly the world could lay waste plans. Their escape from London still weighed on them. In their haste, they had no time even to say good-bye to friends. The ship's staterooms had served as women's dormitories; Charles Jr. slept on a sofa in one of the lounges that had been converted to a men's dorm.

And what place worse than a sun-drenched paradise like Los Angeles to be exiled from the great mission of your generation? Coming from Baltimore, Charles felt LA's tropical luxury like a reproach, a beautiful palm blanket of affluence thrown over layers of fear.

For McManus Sr., a highlight of his visit would be the Santa Anita Racetrack for the year's big race, the Handicap. He planned to bring two Crown Cork associates there: Russell Gowans, the president of Western Stopper, and George Greenan, the sales manager. McManus made the outing a business event: he invited both men and their wives, along with two bottling clients—a Coca-Cola bottler and a Seven Up bottler, plus *their* wives.

Many aspects of married life tested Mary. She thought she knew the many upheavals that marriage would bring: a new name, new ways of being with someone she loved, even strangely changed roles with family and friends. But for a few months in England, she had found herself in a new country, often alone, grappling with thick accents while her new husband spent long days away at work. She relied on other women. Now suddenly she was back in America but in yet another different land, where the sun set in the sea instead of rising from it. Mary realized that she had married a whole family, and not just any family. The McManuses were a tumult of business dinners, tangled personal conversations and silences, unscheduled visits, and the steady, cold suspicion of her mother-in-law, Eva, who had never warmed up to her. The warmth came in friendship with Russell Gowans and his wife, Lorraine, and was a balm to Mary.

The Santa Anita park was just seven years old. That day the track glistened after an unexpected rain. The pale green grandstand, ringed by palms, overflowed with twenty thousand people. Hollywood was there: Bing Crosby and Clark Gable mugged for newsreel cameras. Horse racing seemed even more popular here than back east, if that were possible. Baltimore prized its blueblood Preakness traditions at Pimlico, the country's second-oldest racetrack, as well as at Laurel Park, southwest of the city, which in three decades had grown a working-class audience. Crown Cork and Seal employees had several betting pools that had big payouts for the winners.

As a newcomer competing with the racetracks back east, the Santa Anita Handicap put up $100,000 in prize money, making it the sport's biggest event nationwide. Seabiscuit, America's favorite underdog, had won the Handicap the year before.

For the 1941 Handicap, the betting grew frothy. It was tempting even for neophytes to join in. "Never bet the popular one," McManus told his son. "Always bet the long shots. You might not win for years, but when you win, you win." Charles Jr. held his peace.

That day the program listed a horse named Bay View, a long shot. Bay View had only run two races and had done poorly in both.

"That's the name of the hospital out by the factory," his father said, referring to Baltimore's city hospital (later known as Johns Hopkins Bayview Medical Center). For that reason the old man placed a bet on Bay View. A sentimental gamble; money thrown away.

"That's just the way he is," Junior told Mary. His father considered himself a long shot too. In his youth, after the gunshot had forced him out of school, he had bounced around at the margins. When he got to New York City in the early 1900s to pursue his dream of becoming an inventor, he had nothing but a hunger. Yet from the edge of industry, he had become one of the richest men in Maryland, with a salary of more than $108,000 (around $1.85 million in 2017 dollars), and his company had moved to the heart of the US economy.

Out of the gate, Bay View soon appeared in third position on the outside. He moved steadily ahead, then eked out a slim lead. The margin narrowed going into the last turn. In the final stretch the favorite, Mioland, surged forward. At the finish line the cluster of horses and riders was a blur impossible to make out from the stands. They waited for the announcer.

Bay View, a 58-to-1 long shot, had won.

It was the biggest win in the Handicap's history. Anyone who placed a $2 bet on Bay View won $118.40. McManus had wagered much more.

"I don't know what Dad bet on that horse, but he won," Charles Jr. said. "And he divided the winnings among all the ladies there."

The younger McManus watched the scene stunned. When it became clear that someone had bet on the dark horse, euphoria ignited around them. A crush of people asked, Who? The delay while their group—the men surprised, one woman laughing—waited for McManus to claim his winnings and the payout. They exchanged glances. Russell Gowan's wife, Lorraine, was an opera singer. Maybe she offered a little aria.

Charles Jr. and Mary had been in Los Angeles for months, settling into this strange new place. Leave it to his father to come for a few days and make a big splash.

* * *

In 1941 the spotlight on the cork business was growing hotter, as journalists tracked it as a commodity of increasing national significance. Cork joined the list of "strategic materials" already rationed for the war effort, such as gasoline, shellac, and rubber for tires. (Japan by then controlled most of the world's rubber supply.) The *New York Times* reported on restrictions on French ships traveling to and from North Africa "with strategic materials for the United States," noting cork among them. "Cork in New Bottleneck," another headline punned. The article highlighted a Commerce Department request to restrict cork's commercial use.

European cork suppliers eyed the changes with concern. Nine months after the Baltimore fire, the Portuguese cork industry's newsletter expressed "our fear that the large purchases of cork . . . by the United States would greatly exceed its current consumption." American stockpiling would put the Portuguese exporters at risk "of a sudden interruption." The newsletter reported that "a responsible US source" had explained "the true meaning of the excess purchases" in recent months: "cork is classified as an urgently needed material by the Army and Marine Supply Commission." As a result, the Portuguese faced the risk of a big increase in American demand for cork, "mainly due to isolation, and motivated by the national defense program." At the same time, the Portuguese felt pressure to double and triple their cork exports to Germany.

Meanwhile the mystery of who set the fire at Crown Cork the year before remained unsolved. Government agents were gathering evidence that sabotage networks were active in the United States. The FBI's informant inside the Nazi spy ring led by Fritz Duquesne found the saboteurs were using shortwave radio communiqués from Long Island to send US secrets to Germany. Duquesne told the FBI's mole that he planned to blow up an electric power plant in Schenectady.

The government, growing nervous about the security of critical defense supplies, served McManus with legal papers. The Federal Trade Commission charged Crown Cork, along with Armstrong Cork, with conspiracy to suppress competition. The FTC called out Crown Cork for having half of the world market in manufactured bottle crowns, saying it "hindered and prevented price competition." New federal rules also banned the media from publishing information about cork and other "strategic or critical materials." Besides the FTC, the Commerce Department had concerns that coalesced in a new report that would be released very soon.

"With the coming of mechanized warfare," the FTC said, "cork has become a critical material, one for which few satisfactory substitutes have been found." The report detailed the long-term increase in cork usage from bottle stoppers to a profusion of products in corkboard, insulation for pipes, arms, and engines. The Commerce Department argued that the use of cork had expanded so much that the laws of supply and demand were no longer sufficient; the American cork industry was now a matter of national security and deeply vulnerable to America's enemies. Cork had become an irreplaceable ingredient of modern warfare: it was in everything from bomber plane gaskets and insulation to tanks, submarines, cartridge plugs, and bomb parts.

Yet Nazi Germany undermined America's access to cork. The Commerce Department reported that "the men from the Third Reich have been strongly attracting sellers," scooping up about 40 percent of Spain's cork exports. Even worse, America's cork-dependent industries were all clustered along the East Coast, the region most vulnerable to German sabotage.

"No complete substitute has been found," the authors noted, despite the advent of new materials like fiberboard, mineral wool, and plastics. So the Commerce Department declared, "The cork industry's entire stocks of cork and cork products have been set aside as a reserve ... from which withdrawals can be made, first, for defense needs, then for restricted civilian needs."

All of this weighed on McManus. But the trip to Los Angeles brought a change of scenery and a reunion with his son and daughter-in-law. He saw that his son had the glow and confidence inspired by a life partner. The father felt a twinge of guilt over the England episode—putting the young couple in that danger—but they were here now, safe.

* * *

By then, McManus Sr. had a deep professional interest in California beyond the two factories he owned there. The state had become a detective story for him, holding the solution, he hoped, to his problem of the Nazi blockade.

During a routine business visit to the San Francisco franchise two years before, McManus had taken a drive with Gowans, the Scottish-born plant manager, out to Napa Valley. Beside the road McManus saw a familiar-looking tree and asked Gowans to stop. "That looks like a cork tree," he said.

They stopped the car and got out. The tree was decades old. Gowans took out a penknife and cut off a piece of the bark. It sure felt like cork. But how did it get there? Gowans said he would enlist a horticulturist to check into it.

Soon after that, they were in Palo Alto, walking across the Stanford campus. Another tree made McManus pull up short. He lingered a minute to inspect more closely what his hosts called a "scrub oak." The longer he looked, the more he was convinced that it too was cork. He asked to take some bark and recognized it as excellent-quality cork. On the spot, he made plans to remove the bark from more trees for thorough testing. He ordered his staff to collect acorns to see how well these California trees were locally adapted for growing elsewhere in America.

By the time they got back to the Crown office on Portrero Street, McManus's mind was whirring with possibility. Somehow Spanish colonists had managed to grow cork oak in California many decades earlier. Gowans knew McManus and saw where this was heading: he was going to have to track down every *Quercus suber* in the state.

It was a long shot, but with the Atlantic blockaded, the industry didn't have other good options. McManus asked Giles Cooke to work faster on US climate maps showing zones favorable for growing cork. McManus saw a path for growing his way out of this problem domestically. But that would take years.

* * *

Their investigator on the ground in California was an unlikely sleuth named Woodbridge Metcalf. He was a forestry professor at the University of California at Berkeley, known around campus for his study of firefighting methods and for his teaching at summer camp. Metcalf had a curious nature—he was fascinated by California's native plant life—that led him to the dusty roads and farms where those secrets flourished.

Metcalf was a blend of dreamer and problem solver who had come to California from the Pacific Northwest with his wife. They arrived in Berkeley for their honeymoon on the brink of the Great War, and had stayed and raised a family there. In the two decades since, Metcalf had seen how state forestry officials cozied up to the timber industry, spinning state policy and contracts in favor of the lumber companies. During the Depression, Metcalf saw government relief programs start with good intentions—creating housing in parks while providing jobs, for example—only to yield misguided results. The parks ended up with cabins that were intended to house families but with no toilets or cooking spaces. He was no fan of bureaucracies, but he understood how they worked.

Metcalf had been fascinated with cork oak for decades. A decade before, he had begun a research paper on cork oak in California and had found that the *Quercus suber* plantings dated back to the period of Mexican rule. They also apparently stemmed from a shipment of acorns arranged by the Patent Office in San Francisco in 1858, when ships had become fast enough to beat the cork acorns' expiration problem. Yet the trees had not spread as quickly as planned. In the shift to an Anglo-dominated culture in the years that followed statehood, the knowledge of how to harvest and process cork had been lost by the time the trees were mature. They grew for decades, the cork ripe for peeling, untouched.

The findings from Metcalf's detective work became the core of the McManus Cork Project: to make America free of dependence on foreign supplies; to popularize varieties of cork adapted to the North American climate; to get families and 4-H clubs involved, by shoveling earth for the young seedlings, nurturing the soil; and to create groves of trees to rival the production of the Mediterranean. The project would bring a flourishing of rural businesses for harvesting the trees and feed the industry in an environmentally and economically sustainable way.

At the same time, planting trees would give Americans another way to work through the wartime fear that was brewing.

McManus may have also felt prodded by the public pronouncements of Henning Webb Prentis, his rival at Armstrong. Organized, articulate, and a practiced public speaker, Prentis had stepped up his talk against Roosevelt's economic policies. A few months before, in November 1940, Prentis had joined a radio debate decrying the encroachment of the federal government into business. Prentis complained that the government was usurping economic power; if it had encroached as much on individuals' civil liberties, he said, Americans would have risen up immediately. In a Cleveland speech, Prentis declared that bureaucratic red tape was choking the nation's defense.

McManus was in many ways the opposite of Prentis: not comfortable with public pronouncements, not a writer except for patent applications. The notion of launching a private, grassroots campaign as a response to the war—rather than giving speeches—made growing trees more appealing to the taciturn McManus. Even if the domestic planting effort never fully replaced cork imports, it would be worthwhile. On the other hand, if it failed, the growing project could tarnish Crown Cork and strike people as an oddball waste of money.

With a project to adapt cork to the American climate, McManus became convinced he could chart new territory. California could be the next frontier in that. His first inventions, back in his early New York days, had expanded the uses for cork; maybe this phase of innovation would expand something else—the world's natural supply and people's appreciation of the material. Here in America where nobody thought it could grow. A long shot.

* * *

Junior held a slightly more cautious view. It wasn't that he disliked Cooke, who was a fatherly kind of guy, honest and dedicated. People trusted him and listened to him, and he was methodical as hell. He was just the person who could make such a scheme work. Who else in the entire country had written a PhD dissertation on cork? But Charles Jr. wasn't sure what to make of Metcalf, and whether or not his survey of California for cork trees was a wild goose chase.

One day when Metcalf was in Los Angeles, Charles Jr. took a few hours off from the logistical headaches of the new factory and of nursing relationships with clients and went to see the cork tree operations. The idea of surveying all the oak trees across the Golden State struck Junior as the kind of publicity stunt that his father normally hated. Yet here they were, doling out tens of thousands of dollars on a slim chance of finding a homegrown solution to the cork shortage, an effort that wouldn't yield any significant results for a decade. Junior got in the car with a sigh.

Driving west in the afternoon light, he saw that the city had a sort of magic, with the light coming off the ocean past the fishing villages, where boats bore Portuguese and Italian names.

In Ventura, standing by the road where Metcalf's crew had just finished, McManus was amazed to see stripped oak trunks shining raw in the late afternoon light, a light red to pink. He knew from his trip to Portugal that in the coming weeks, the trunks would slowly darken, becoming almost black, as if their skin were toughening.

These trees didn't look exactly like the twisted giants he had seen outside Lisbon, but there was a resemblance. Standing near a different ocean now, with the sun sinking down, Charles felt the world lumbering like a great churning engine, spinning, on the verge of seizing up. In a few weeks he would be standing beside his younger brother on Walter's wedding day, back east, bearing witness to a ceremony full of hope against a background of devastating news everywhere. Britain teetered under the pummeling bombardment. In Asia the Japanese were grabbing Indochina. Yet Charles and Mary were talking about having kids. Against all the strangeness, this was not a time to put things off. This was a time to take a breath and plunge ahead.

* * *

In his office on the Berkeley campus, Woodbridge Metcalf assembled an odd assortment of tools: a back-bending saw, a Portuguese ax, a wooden mallet, and an item fashioned from a car's spring leaf—a compound coil of metal strips, essentially a rib spreader for a tree.

Metcalf had designed some of the tools himself. When he couldn't find anyone who had ever stripped a cork oak, he went to visit a saw manufacturer in Chico, a three hours' drive north, past Yuba City.

"You make pruning saws with the teeth on the inside of the curve," Metcalf said to the man. "Do you think you could make one with the teeth on the outside?"

The saw maker had never seen such a creature, but why not?

That specially adapted saw became the contraption Metcalf used to make the first vertical cut through the cork layer, to avoid damaging the inner cambium. It cut like a scalpel's thin line to the pericardium before the rib spreader took over. The saw was more delicate than the Portuguese cork-stripping ax Metcalf had. He also tried a conventional hemlock stripper he had acquired back east, but like the Portuguese ax, it scarred the tree's living cambium. The Chico saw-maker's version did better.

A wooden mallet was handy once the tree peeling got started. Metcalf liked to describe how the ax handles got the cork cracking away from the trunk "like a ripe watermelon." These peeling operations and the analysis they enabled would help in the next phase: setting up plantations of seedlings around the state "to provide in the United States at least a part of the nation's cork requirement." This was serious. Finally, after years of tracking the odd trees in the landscape, he had someone willing to invest in them. With Crown Cork and Seal's bankroll, the project got a quick start—fifteen seedlings planted that year, with fifty thousand for 1942.

By the time McManus Sr. visited Los Angeles for the Santa Anita Handicap, Metcalf's cooperative cork oak program was in full swing, enlisting his contacts across the state: the California Division of Forestry and the state's Forest and Range Experiment Station, along with Western Stopper. They started combing the state for cork oaks and rumors of cork.

The oddness of cork's interrupted history in California—and resuming a tradition of harvests that had been lost in crossing the ocean—tickled Metcalf. When he came across clues to the first Spanish plantings, he scribbled them in a little spiral-bound notebook with lined pages. "Emory E. Smith," he wrote at one spot, "cut corks and displayed them at Chicago [World's] Fair in 1890 at White City." The cork had come from trees north of San Francisco; other trees grew in Golden Gate Park. "Corks were all cut of various sizes," he wrote. Oldtimer Emory Smith told Metcalf of visiting Spain, where he'd seen a thousand people working a cork harvest near Gibraltar. Metcalf noted that Smith kept a file on the history of cork oaks in Europe at his office at 651 Howard Street in San Francisco.

Another day Metcalf recorded in his daybook a visit to the J. T. Kiser ranch near Sonoma: "Kiser says police chief of Burlinghame knows history of his trees." In the bottom margin, a quote of the day from Dryden whispered, "They can conquer who believe they can."

Once the survey of California's cork trees was complete, Giles Cooke's team would test the samples with chemical analysis and assess ways to propagate the trees that produced the best cork. Then Metcalf would gather acorns from those individuals. The whole operation meant an army of untested foresters spreading out across backyards, hospital grounds, schoolyards, and parks across the state, using the sharpened pikelike weapons in Metcalf's office. In its first fall, Metcalf's operation stripped 248 trees of various sizes from Napa down to Los Angeles. The project amassed more than 10,500 pounds of cork: Five tons of cork in a single harvest by first-timers.

* * *

Fears of war and foreign influences were escalating. California was coming down harder than most states on immigrants labeled as potential enemy aliens. Attorney General Earl Warren took a hard-line stance that most Californians approved. Within months, Japanese and Italian American families living within five miles of the Pacific coast would be forced to relocate, as officials considered them a risk to maritime security. The forced migrations came when finding other housing options was especially hard due to the wartime economy's jobs boom on the West Coast. A group of Italian Americans from a neighborhood forced to relocate from Pittsburg, California, to further inland, took the train to Washington, DC, and lodged a protest against this practice, but the relocations did not stop. The anxious suffering these families experienced was grueling. Some Italian Americans targeted in the Bay Area killed themselves from the humiliation. In Richmond, California, sixty-five-year-old Martini Battistessa offered a friend fifty dollars to shoot him in the head. When the friend refused, Battistessa threw himself in front of a train. A fifty-seven-year-old man, when officials told him he had to leave his Vallejo home, cut his throat. A third jumped to his death from a building rather than comply with the Justice Department's order to relocate.

Back east, Highlandtown families read in the news that in California, "aliens who have been adjudged dangerous or potentially dangerous already have been rounded up and placed in concentration camps under the regime of the War Department." The *Baltimore Sun* noted that the US attorney general had three choices in those cases: release of aliens deemed "not dangerous or unfriendly," parole, or internment "for the duration." The army held detainees under military guard and fed them rations while it built permanent concentration camps.

The timing of these measures against immigrant families surely complicated the emotional responses that Metcalf's cork investigators stirred up among California residents, especially in Japanese and Italian neighborhoods near the coast. Metcalf did little to prepare them for his project's visit or what would happen to their trees. No advance letter; no public awareness campaign. If his team spotted a big cork oak in a front yard, he would approach the driveway or knock on the front door. When the resident opened up, Metcalf would introduce himself and say, "Now, we would very much like to strip this tree."

He then explained the cork experiment and the odds of success. "We can't guarantee that it won't hurt the tree," he'd say, "but we don't think it will. We haven't lost any trees yet," adding, "but there's always a possibility."

Most people said yes out of a sense of patriotic duty. Many wanted to watch the crew do its work.

Why doesn't stripping the cork kill the tree? people asked. Metcalf explained, "The cork oak is the only tree in the world that grows an annual ring in the cork, as well as in the wood." How often could you harvest? Every eight to ten years, he said.

In California's climate, the wood cambium grew from the spring until early July. Then the outer cork cambium started to grow. When the cork cambium was in full growth and the new cells were soft, the crews could safely harvest. They made a vertical cut along the trunk for up to eight feet, then horizontal cuts through the bark. Some observers gasped when the whole shell was lifted off.

* * *

In Los Angeles, a parks and recreation crew in the little-known office of the city forester found around two thousand cork oaks growing in the county. In 1941 the Chandleresque-named Averill Barton with the LA Forestry Department assembled a file with a card on every tree they found, from beautiful spreading canopies along highways to younger trees on Devonshire Boulevard in the San Fernando Valley. The team pored through old plant catalogs from commercial nurseries such as

Theodore Payne's at 345 South Main Street, which in turn pointed them to other plantings.

"The cork oak on the lawn of C. H. Powers, 230 Citrus Avenue, Azusa, is 54 years old, 65 feet tall with a clear length of over 20 feet," Metcalf recorded. He added, "It is one of the most beautiful cork oaks in Los Angeles County."

The city's sweet spot for *Quercus suber* was between Chatsworth and Zelzah at the valley's western edge, where orange and fig trees grew. In the future, it would become a crucial wildlife corridor. Near the ocean they found a scattering of cork oaks in Ventura County, including one in a citrus grove belonging to R. H. Peyton near Rancho Sespe at Fillmore, a resort for trout fishing. With a crown spreading sixty feet wide, it towered above the orange trees. Metcalf's team stripped two hundred pounds of cork from its flanks.

Finally, a few miles east of Santa Anita in Duarte, a hub of citrus and avocado growers (the town was part of an 1841 land grant from the Mexican governor to Andreas Duarte), they discovered a cluster of thirty-year-old cork oaks. Some trunks were more than twenty inches around, about the size of an underfed teenager's chest.

With their axes and knives, Metcalf's team sliced away as Hollywood-bound traffic whooshed past and residents watched from their windows. The team thrust their knives into the cork shell and prized off eight-foot-long slabs, one after another.

Metcalf bound up the slabs and packed them off to Baltimore for testing. There, Cooke became intensely interested by what he found. The *Baltimore Sun* reported that Crown Cork was testing California-grown cork, reminding readers of the severed supply chain from Portugal and that cork was crucially important for its "use as washers and gaskets in the engines of trucks, tanks, airplanes and the numerous power devices that go to make up the modern fighting machine.... It is pulverized for the manufacture of insulation for buildings, refrigerators and ships. It goes into the soles of shoes, life preservers, floats to register the quantity of liquids of tanks, bobbers for fish lines, tips for cigarettes, handle grips and in cartridges and bombs."

The last place Metcalf traipsed around with his bark-ripping tools was the Napa State Hospital for the Insane, in the countryside on the Napa Vallejo Highway. About an hour north of Berkeley, the hospital was a confusion of architectural styles dating back to 1875, with more than three thousand patients. With war coming, patients showed their patriotism by making camouflage netting and stretching it out on the lawn, unaware that a valuable part of the war effort already grew on the grounds. On Metcalf's first visit with research associate Palmer Stockwell, they inspected three large cork oaks and several small ones. Metcalf noted that the hospital had some of the most spectacular cork trees he had ever seen.

He spoke with the grounds department manager and left a message for the hospital manager and campus engineer. "We wanted to get all acorns possible in December," he wrote. "All trees we saw are evidently going to have a good crop of acorns."

Metcalf returned to the Napa hospital many times, becoming familiar with the grounds staff and occasionally putting on an unexpected show for the residents. One morning he and his crew of half a dozen men, including several sailors, filed past security guards into the locked facility, toting sharp axes and large ladders. This caused a stir among some of the patients while the crew laid out their equipment in the shade of one of the larger trees.

They made their preparations, checking the equipment with some nervous energy as they considered the age of the tree they were about to slice into. They didn't know for sure if they were advancing research that would find the best-adapted versions of this species or if they were mortally wounding some of the last local giants. Hospital staff and patients looked on as the foresters hoisted the axes and started to slice through the bark, careful to feel where the pressure increased at the inner cambium, and an observer might have found it hard to pick out the performers from the audience. Where was the line separating outlandish spectacle from real life?

Thirty years later, Metcalf recalled those visits to the hospital, with the same odd cork-stripping tools still hanging close by him on his wall, distracting the attention of a visitor. "The largest cork oak in the United States, as far as I know," Metcalf told the visitor, "is on the Napa State Hospital grounds. That tree was planted in 1873 on very good land."

A few months later, Metcalf wrote in his daybook: "Office in A.M. and dictated a lot of letters to Mrs. Gatewood. Got a small camera from Ernst and a tripod and met George Greenan at the garage at noon. We had a light lunch at home and drove to Napa State Hosp. where we took pictures and found the acorns starting to ripen on the trees there. Also called at Mrs. M. Armstrong's place on Foothill Blvd., Napa and saw a 28.8" tree which she does not want stripped.... Leave by 8 PM."

These visits fell within the orbit of a forester's daily life but were subtly different. People across the country experienced those months of 1941, when America seesawed toward war, with varying degrees of fear and resolve. Metcalf worked with the Red Cross on preparations for handling wartime emergencies up and down the coast. But nobody really knew what was coming, from the families in the fishing villages and the jockeys at Santa Anita to the old-timers in the hills and the inmates at Napa State Hospital.

* * *

Charles Jr. came away impressed by the cork-harvesting efforts. With the war, many things were changing in American life and business. Maybe Americans were eager for an active role in growing the substance of their material lives and national security. Still, the cork-growing project, like Crown Cork's defense contracts, seemed less strategic than Armstrong Cork's approach. Crown Cork's business still rested squarely on the low-tech *Quercus suber* tree, its tough shell, and its loose militia of rural harvesters scattered around the Mediterranean's edges. Experts like Cooke said that the bottling industry's reliance on cork wouldn't change after the war—consumers would continue to demand cork-lined containers due to chemical reactions to carbonation and taste preferences.

Charles and Mary remained in Los Angeles that spring, until the LA factory was up and running. By the fall they were home in Baltimore, where the Highlandtown plant was back nearly to full operation. Junior was once again commuting to Highlandtown every day. Crown Cork and Seal made new, modern products, but the cork yard at the Baltimore factory felt ancient, with acres of mountains of bales.

The war was stirring fears and shifts in US policy positions, which made Junior uneasy. In October 1941 the government released its disturbing report in *Commerce Weekly* with a boldface title: "Cork Goes to War." The cover of one edition showed a map of the Mediterranean with the lands where cork forests grew: Portugal, France, Italy, Spain, Algeria, Morocco. A tightening noose represented the Nazi blockade.

Charles leafed through the report somewhat embarrassed. In a way, it reflected how well Crown Cork had expanded through the previous decade: imports to the United States had increased to nearly 4 million pounds monthly. Around two-thirds came from Portugal; just under one-quarter came from Algeria. His father had consolidated the largest corporation of its kind in the world. With his success—and the success of Armstrong Cork—came this unwanted government attention. This was not just because cork factories were vulnerable to sabotage. They also posed another vulnerability: cork was an industry where a few players dominated the market; they could leverage that position to extort America.

* * *

For decades, cork's pliable functionality and resilience had inspired McManus. Now the McManus Cork Project was weaving an even tighter bond among the communities of cork, from 4-H kids across the American South to herders in cork forests of northern Africa, to secret government agents. The industrial process capitalized on cork's adaptability—a blossoming of inventions that exploited its compressibility, its sealant powers, and more. And the trade from cork harvesters in Europe and North Africa had grown to the massive scale noted at the World's Fair, at the Paris Exposition, and in the report "Cork Goes to War."

Three weeks after Pearl Harbor, Crown Cork and Seal executives traveled to Washington and sat down with officials at the War Department for a classified meeting. According to official notes, the aim was to explore how the Defense Plant Corporation could purchase the company's downtown Guilford Avenue factory (Crown's original location from the days of William Painter) for wartime manufacturing. The Defense Plant Corporation, a subsidiary of Roosevelt's Reconstruction Finance Corporation that found alternative sources for essential materials (for example, it helped to fund the creation of synthetic rubber), issued contracts for everything from bearings to bomber planes. Roughly half of the agency's budget went to aviation. At the meeting, David Stone, the general manager of Crown Cork's machinery division, described an existing Crown contract with the government for making "tripods," or machine-gun mounts. Stone explained the Guilford Avenue plant setup—its passageway connections and security limitations. The new contract would add a five-story building next door and expand the old plant so that Crown could increase its tripod production to ten thousand a month, a more than tenfold increase.

On a sheet of scratch paper, Stone sketched floor space calculations, a layout for the machinery, and a ballpark cost of the new defense works. After the meeting he and the other Crown Cork reps left Washington with a deal with the War Department. The government quickly contracted with Armstrong Cork too. These and other wartime contracts would help the United States meet Roosevelt's mind-boggling industrial goals: sixty thousand aircraft in 1942 and twice that number for 1943.

Within a year, Crown Cork had a growing portfolio of defense contracts, including its subcontract to build bomber wings in Highlandtown for Glenn L. Martin and the B-26 Marauder.

* * *

In 1942 McManus's campaign to grow cork oaks across America gained momentum. It was perfectly timed for government support. For the McManus Cork Project, a military transport plane left Morocco carrying acorns collected in the Atlas Mountains, specially chosen to resist North American winters. Nursery workers then planted the acorns in a greenhouse at the University of Maryland—a modest piece of the national security infrastructure. Giles Cooke supervised the operation and conducted chemical analyses downtown at the Crown Cork lab, mapping a crescent of climate zones where cork oak could grow successfully across the country's southern rim and charting which varieties would grow best in each zone.

Baltimore threw itself into the war effort. Baltimoreans recycled everything for national defense—old tires, scrap iron; even old record collections were sacrificed for their shellac, a precursor to vinyl. One young Baltimorean, John McGrain, watched his uncle's record collection fly out the attic window, disk by disk. Then he and his brother collected them outside and hauled the shellac pieces patriotically away.

The cork-growing campaign ramped up to distribute 7,500 pounds of acorns a year nationwide. (At its peak, the project would send out nearly twice that.) Crown mailed out acorns and seedlings with instructions on planting methods, potential insect problems, and watering suggestions, all illustrated. McGrain, the boy who tossed his uncle's records out the attic window, read in the newspaper about a cork tree–growing project. It appealed to him in the same way that science projects drew him, and he mailed off a request for acorns.

By return mail came a box packed with damp moss. The package was soggy when McGrain opened it and found in the moss three acorns. He planted them in the yard and watered them, and after nearly a week they sprouted. They kept growing and became shrubby. He watered them until they grew up to twelve feet high. The seedlings had surprisingly dark leaves, he noticed. Dark, dark green.

So many things turned out to be not as they seemed—the war factories around the city were shrouded in camouflage; even trucks hauling wings for bomber planes were camouflaged—that it struck the boy as odd to get something as simple and tangible as a box of mossy acorns. Planting them for America was a small, concrete action that he could take.

Other Americans—including many immigrant families—were playing their part for their country overseas. Calls to patriotism led the Marsas and the Ginsburgs into murky situations more dangerous than planting trees.

Chapter Five

# SERVING THE CROWN IN WARTIME PORTUGAL

## · Marsa: 1940–1942 ·

> One chooses for peeling or lifting cork the summer, from mid-June to the end of August, thus avoiding the moment of full activity of the sap, the rainy season, the dry hot winds, at the end of which the mother layer, gorged with liquid and made up of new tissue, is not vulnerable to being torn off nor surprised by bad weather . . .
>
> A. MATTHIEU, *Algerie: Flore Forestiere*, 1897

Gloria Marsa was preparing for her second evacuation from Europe, this time from Portugal. She loved Lisbon's picturesque hills, and the apartment near the Parque Fernando Septimo, a large monument to the Marquis of Pombal. Septimo had been Portugal's secretary of state in the 1700s who had responded swiftly with relief for the survivors of the disastrous earthquake of 1755.

Everything in Lisbon made history intensely vivid. It was a colorful, bustling city, a metropolis that coupled the modern with the old, more cosmopolitan than Seville had ever been. Perched over the wide Tagus, Lisbon flooded with a light as spectacular as that of Naples or Sicily. Flowers dotted the cityscape with red and orange accents against the blue background.

"Lisbon is built on seven hills," Gloria liked to say. "And you're forever climbing either up or coming down. But you're never on a street flat level." She liked when taxi drivers complained about the street where her school stood, saying, "It's very *violent*" because it was extremely steep.

Some mornings, if she was running late, the view out the taxi window down the hill to the water was her last exhilarating moment before going to class.

The Marsas had lived in Lisbon for more than two years. Gloria was there for the Portuguese World Exhibition, celebrating eight centuries of Portugal's culture and independence from Spain. The exhibition, a response to the New York World's Fair, lasted for six of Europe's most tumultuous months, from June to December 1940, and presented to the world António de Oliveira Salazar's vision of Portugal's superiority. Salazar had a new marina built just for the event, along with a replica of a seventeenth-century galleon.

Gloria's father worked feverishly for Crown Cork, building a new plant for the company's Portugal operation. Portugal was one of Europe's few neutral trade terminals, still open to ships from the United States, South America, and Africa. Lisbon newsstands sold papers and magazines in all languages. Cafés and bars swarmed with conversations in many dialects. Gloria loved that you could feel the entire world just by walking down Lisbon's streets.

Along with the cosmopolitan atmosphere was the palpable air of fear. The city was jammed with hundreds of thousands of refugees, all fleeing a Europe under siege. They came to Lisbon, hoping to escape to the United States, leaving behind ravaged places, shattered lives in French camps, family members deported to Poland, the work camps. There were rumors of extermination camps too, but few people believed that the Germans would divert energy away from their military campaign. The rumors jumbled together and created radically opposing visions of what life might look like after the war.

Salazar ruled Portugal with a clenched fist. He had built the country's economy and allowed industry to flourish while repressing internal dissent and any sign of communist activity. During Spain's civil war, his police delivered Spanish refugees suspected of being communists back across the border to waiting Spanish police and certain doom.

At the same time, Salazar showed sympathy to the new refugees pouring into Portugal as he had declared that no hotel, restaurant, or service provider could raise its prices to take advantage of them. But it was unclear how long his government would stay neutral, as Salazar could bend to external pressures. Portuguese scholar Augusto D'Esaguy published a slender book titled *Europe 1939*, including descriptions of how German Jews were robbed of nearly all their belongings as they emigrated and of Germany's brutal treatment of Polish Jews. When the book drew a strong protest from German diplomats in Lisbon, Salazar ensured it quickly disappeared from bookshops in Portugal.

Salazar's forces treated foreigners carefully to avoid international incidents. But there was a definite pattern in the treatment meted out by his Surveillance and State Defense Agency, known as the PVDE. It was harsher when the Axis Powers were winning. In 1941 a train with Jewish refugees arrived from the Netherlands but was turned back at the Spanish border, into an uncertain future. The Axis sometimes flexed its muscle even inside Portugal's borders. In one instance that year, the PVDE abetted German agents in kidnapping from Lisbon's streets the German Jewish journalist Berthold Jacob. (He had exposed Germany's secret rearmament in the 1930s and other crimes.) From Lisbon Jacob was hauled away to prison in Germany and died three years later.

* * *

The Marsas received a steady flow of visitors during this period. One group was the McManuses: Charles and Eva came with their son Walter, who was a few years older than Gloria. The two families had dinner together, and Charles combined work and pleasure, assessing Crown Cork's navigation of Portuguese bureaucracy while exposing his younger son to the world. The McManuses, who always impressed Gloria as gracious, brought the Marsas up to date on the latest news from New York.

That summer after Germany invaded Belgium, the Marsas received Crown Cork's manager for Brussels, Antoine Leenaards, and his family. They were among the thousands fleeing the Nazi aggression, hoping to reach the United States. But border crossings had become much harder to navigate. In Spain the Leenaards waited weeks for papers so they could continue to Portugal. Finally, in August, Gloria's father managed to obtain the permits. He traveled to Spain to bring Leenaards and his wife and son to Lisbon. The days slipped by still with no sign of Melchor or the Leenaards. Pilar Marsa accepted that her husband would miss their wedding anniversary dinner on the twenty-seventh. But at the last moment on that Tuesday, he walked in the door with the Leenaards family.

The dinner at the Marsa home that night was a moment of relief and celebration. Amid the fall of France, at least their friends had escaped. The tumultuous times caused Marsa and other corporate leaders in Portugal to reevaluate how they used their positions with foreign companies. Marsa mainly felt an allegiance to his colleagues and his employees. Every year he hosted a holiday dinner for his Lisbon staff at the Hotel Aviz.

Marsa respected how the professor-turned-politician Salazar had kept Portugal out of the war, despite pressures from both sides. More and more refugees continued to pour through Lisbon, so many that encampments grew up east of the city. "Every European nation, religion, party was represented in that procession," wrote author Arthur Koestler, one of the refugees who crowded the city in those months. Many were Jewish refugees who had fled Paris. All lived in limbo, waiting for visas, travel arrangements, or money from relatives. And all feared the PVDE, which tracked the exiles closely throughout their time in Portugal. The Portuguese regarded refugees warily, and officials kept tight restrictions on their ability to get jobs or receive state benefits.

That winter Marsa realized the situation had become too dangerous. The Nazi blockade threatened to cut off all exit to America, so the Marsas prepared to flee for safety. On January 10, 1941, they boarded what many believed was the last ship out of Europe.

Gloria stood on the deck of the *Excambion* once again, looking up at the city and marveling at it. She exchanged words with another passenger, a foreign correspondent who had recently fled Paris. "How will the continent look when we return?" they wondered aloud.

Both sides of the ship's hull bore a huge painted American flag—brightly lit at night. Gloria thought, "This is a sure target for the torpedoes. We're in everybody's sights." Maybe the captain, realizing that Germany did not want the United States to enter the war, was advertising the relationship to head off U-boats. Still, it was a risky bet. Throughout the ten-day journey, Gloria recalled later, the passengers asked each other, "Will we arrive?" It was a haunting voyage.

* * *

As the family settled back into life in Brooklyn that spring of 1941, Gloria saw that New Yorkers' main fear was getting drawn into the European conflict. Conversations about politics were more frightening than ever. Americans were deeply divided on whether the United States should enter the war. Some argued that America should let Germany and Russia fight it out, exhaust each other, and then America would swoop in and establish peace. The America First Committee, which by then had 800,000 members, held a similar view of disengagement. Charles Lindbergh gave speech after speech, declaring that the war "was not our fight, that we should let the others do it and meanwhile strengthen the country."

That September a fire destroyed the cargo being offloaded from the freighter *Sines*, docked in New York Harbor. The *Sines* had come from Portugal with 300,000 pounds of cork, now scarcer than ever. With the censorship rules governing reports about cork, newspapers published only a short paragraph about the disaster, saying that the fire had burned for about an hour. "No estimate of damage was made," the *Times* reported. "No evidence of sabotage was found."

But sabotage was on people's minds. Just three months before, in June 1941, the FBI had arrested a wasp's nest of Nazi spies based in New York. J. Edgar Hoover was jubilant that the FBI mole among Fritz Duquesne's ring of spies had yielded twenty-nine arrests across four states. Hoover declared the operation "the greatest of its kind in the nation's history." In newsreels, Hoover showed German spies with blueprints of steamships vulnerable to bombs and described in chilling detail a wide-ranging network of saboteurs ready to blow up important industrial sites in the United States.

One evening in early December, Gloria was visiting a friend's home. As they listened to the radio, the seven o'clock news came on and reported that Pearl Harbor had been devastated by a surprise Japanese attack. Everyone in the room was stunned and silent.

* * *

Port security in New York tightened as in Baltimore, and the Coast Guard's volunteer patrols kept watch on the comings and goings of freighters. Under the new rules of silence and secrecy, newspapers could not publish even statistics about America's cork imports and reserves, now classified as secret.

In the spotlight on waterfront activity, the focus was sharpest on ships from Portugal. Their crews worked under intense political pressures. The Portuguese freighter *Pero de Alenquer* made frequent Atlantic crossings. On some voyages, like the one it made in November 1942, the *Pero de Alenquer* carried hundreds of Jewish refugees fleeing Nazi-held France. Other times it carried cork to New York and Baltimore. Often crews had divided loyalties: some sailors had worked for the Spanish Republican side; others were underground supporters of Salazar's opposition. So the ships were ideological powder kegs.

At least once, in July 1942, Nazi blockade authorities in Lisbon ordered Portuguese shipping companies to unload cargo only in specified ports, diverting New York cargo to Baltimore. This was sheer intimidation—the German officials said that the crews had to obey or "sail at their own risk." As a result, Portuguese shippers said, "We shall not be able to ship more cork because Baltimore is congested and unable to handle" more, reported the *Washington Post*.

Portugal had the full attention of Washington officials. US strategists were worried that the Iberian Peninsula's trade would tip the war's outcome. With North Africa held by the Axis Powers and Spain still weakened from civil war, Hitler could take the peninsula and close the Mediterranean to the outside world. A German occupation of Spain was a very real danger and a strategic risk to the Allies. And Portugal was a principal source of tungsten (then known as wolfram), crucial to Germany's arms industry.

In this situation, economic intelligence was often the greatest prize. Not wanting to be caught flatfooted by his new agency rival, the OSS, Hoover kept a flurry of classified memos going to the OSS director in early 1942. His memos assailed Col. Bill Donovan with a raft of economic news on everything from Portuguese tinplate (Germany had offered to buy all Portugal's production) to platinum (informants reported contraband platinum smuggled into Lisbon, allegedly from the United States) and tungsten exports. Hoover even sent memos about Portuguese imports of animal hides from Brazil.

The OSS had its own strategies for tracking shifts in commerce. Following the lead of British Intelligence, the agency set up ship observer programs, with listening posts in major ports. "The ships of a neutral country are carefully watched and the sailors tailed when they take shore leave," explained OSS agent Donald Downes. "One or two sailors are discovered to have cousins, uncles, or sweethearts in the city." Once investigators learned that information, they then found out whether the sailor's contact was a US friend or enemy.

Among Spanish and Portuguese crews, the OSS often found sailors sympathetic to the Allied cause through exiled political groups, particularly the Basques. In the wake of Spain's civil war, Franco's Nationalists executed more than twenty thousand Republican sympathizers, including Basques; many who escaped were willing to help fight the fascist regimes for revenge. A ship's crew member befriended by an OSS agent "could observe enemy vessels under repairs; content of cargoes and their destination; bomb damage; civilian morale" and more. In a crisis he could deliver a message or a small radio receiver. And if the sympathetic sailor could manage to get time alone in his ship's radio room, he might discover if the captain had betrayed Allied ship locations to the enemy or independent raiders. Ship observers provided proof that Spain was refueling Axis submarines with oil from Allied sources—"oil granted as appeasement to Franco, and under his solemn promise not to allow a drop to fall into enemy hands."

The *Thetis*, a Greek ship, routinely plied the sea from Lisbon to Baltimore. It was loading wheat at the Western Maryland Elevator Pier in Baltimore in mid-1942 when OSS agents spoke with the crew members. Most of them were Greek except for one Portuguese crewmate and a Spaniard named José Robert, who worked in the engine room. Robert was an anarchist and an exile from Franco's Spain, unable to get a Spanish passport since the civil war. An OSS agent identified only as S. described Robert as a quiet man, slender, middle-aged with thinning brown hair. The right side of his face was paralyzed, giving him "the physiognomy of a terror film character," but one with a mild manner. Robert said he just wanted to help the underground Spanish Republicans working against Franco.

The OSS used men like Robert to infiltrate crews hauling cork and other commodities and persuaded them to give information for the Allied cause. Agent S. wrote, "They all will be convinced that our men in Lisbon are working for the Spanish and anti-fascist European movement."

"I asked him [Robert] to get in touch with our men in Lisbon," S. reported. "He was convinced that the man he will contact has something to do with the Spanish underground movement."

The *Thetis* left Baltimore the next day for the two-week crossing to Lisbon. As soon as it docked there, José Robert made his way to Corpo Santo, a square near the riverfront dominated by a Baroque church. At the Café da Bento he waited, expecting to meet his OSS contact at around 6:00 p.m. The contact gave the name Ricardo as a password.

Robert produced from his pocket one half of a postcard. The agent showed the sailor the other half of the postcard. They talked over drinks. Robert told what he knew about boats going between Lisbon and Genoa, stopping at Gibraltar. And he took a message from the agent to give to another contact.

OSS agents befriended sailors in New York's East River as well, tracking boats bringing refugees from Europe and making contacts among shipping agents near the Battery. In June 1942, agents observed the Portuguese ship *San Miguel* as it was loaded at Pier 28: "Noted crates marked for Portuguese Air Ministry in Lisbon. Also pressure tanks of the type used for oxygen or liquid air. To get on the dock, one has to get through police and coast guard." They tracked the *San Miguel*, a boat that could hold only twelve to fifteen passengers and usually took only two or three. It left New York on June 12. The main cargoes were going to Baltimore. A July 20 OSS memo read, "It seems sure that from now on, all our boats are going to arrive in Baltimore (under pressure from the German naval authorities, who seek a technical blockade of the Port of New York . . .) so some of the activities dedicated in securing my entrance to boats in New York must be repeated for access to same in Baltimore."

In Baltimore, Agent S. noted the inroads that spies were making among Spanish crews. Agents had to find ways to speak with sailors away from their crewmates, since many were Franco sympathizers. He said they needed to "find the good guys who work on the pier, and see how sympathetic the longshoremen are." He also noted a rise in blackmail by German agents against officers on Portuguese ships. American spies needed to exert the same "toughness and pressure," he wrote. S. hoped to use "Spanish friends" like José Robert to make headway infiltrating the Portuguese crews.

One July afternoon, Agent S. stepped aboard the *Pero de Alenquer* and was escorted into the captain's cabin. The captain cleared out the crew members so the two men could talk privately. Above them hung a portrait of FDR and photos of the captain's two sons. Captain Gomez explained his situation: he was sympathetic to the Allies, he said. But he feared getting involved. Detention by the notorious Portuguese police was a hideous prospect, he said. Even worse was the effect that his arrest would have on his sons.

Gomez offered the agent a glass of port and a few names, and he suggested that the agent contact his boss if he wanted more. Captain Gomez would only provide what he was ordered to. He wished the visitor a good day. S. called on the captain again a few days later and left frustrated by the man's reticence. He wrote in his report, "a little implied threat is needed to clinch cooperation."

A little blackmail.

* * *

Agent Donald Downes was a significant figure in the OSS and its East Coast operation. Downes, Yale class of 1930 (a credential helpful for espionage), read voraciously and joined arguments with gusto. After graduating, he had worked as a schoolteacher in New England for several years. Then he applied to the Office of Naval Intelligence. He read a book called *The Strategy of Terror*, "which described how Hitler was deploying a fifth column through all the democratic world." Downes reread the book, finding that it explained profoundly things he had seen on his recent trips to Europe.

Downes took his first spy job with British Naval Intelligence, with a teaching position in Istanbul as cover. He was out of step with American public opinion. At that point in 1940, four out of five Americans opposed war with Hitler. In helping British-led spying operations on both sides of the Atlantic and not informing US authorities of his role, Downes risked being charged with treason when he returned to the United States in early 1941.

Then Pearl Harbor changed the alignments. Suddenly, with America joining the Allies, Downes was on the right side of the law. He brought his contacts with him to the new US spy agency and moved up in the OSS ranks. One of his contacts was a Yale friend named Robert Ullman, who worked for the news agency Pathé. Ullman offered Downes an office inside his own, accessible day or night. It provided "maximum security against a search by FBI, or worse, some enemy organization." Downes told his friend, "It's important secret work for the government and you mustn't mention my presence to anyone."

Downes had expansive ideas about what intelligence could do for the war in Europe. Much more than just gathering information for military strategy, OSS informants had the potential to help overthrow enemy governments and even neutral regimes like Spain and Portugal. He pressed his OSS colleague Allen Dulles in a classified memo in July 1942: "Why do we not take proper steps to get into the hands of our millions of allies in Europe the materials of destruction? . . . The contacts exist." He continued, "If we could . . . supply the organizations willing and anxious to set a prairie-fire of resistance in Europe, we could, I feel, get far greater results more quickly than any other way." At the end of the letter, Downes came to the point: he recommended the establishment of a special section of the OSS that would provide rebel organizations with the tools for a general uprising in Spain, "even possibly with a full-dress revolution."

A spy's life in New York at that time looked a lot like a publishing career: a room rented on the Upper East Side on a quiet lane off East Fiftieth Street under the auspices of Pathé News; expense receipts for lunches and taxis to Penn Station (with suitcase), a ticket for the night train to Washington (lower berth), breakfast (fifty cents), hardware (fifteen cents), phone, tips. Wrangles over unused bus tickets to Long Island. Memos confirming appointments, including a sit-down with the agency's director, Colonel Donovan, to discuss "the Foreign Nationalities problem." At one point the OSS even placed an ad in the *New York Times* for a photo contest, inviting snapshots of military installations in Europe, with the promise of a spread in *Life* magazine.

Besides the office on Madison Avenue, Downes had a Maine address in Yarmouth, up the coast from Portland. When he traveled to Algiers, he received his *Time* magazine in care of the Allied commander there. His job involved writing letters, for example, requesting information about alleged Gestapo agents operating under cover of the German Consulate. Downes arranged for OSS informants to get passage to Europe; sometimes they shipped on merchant ships as sailors and were given $200 each to reach a rendezvous point, where they would jump ship and await his instructions and monthly pay for their services onward (at a rate comparable to a US Army captain's salary).

Downes had a special interest in getting the shadow government of Spain's exiled Republicans on board with US spy operations, and then training them to enter Spain, Portugal, France, and North Africa. He also helped with OSS efforts within the United States to expose Italian American sympathizers of Mussolini. From his friends in Italian enclaves along the East Coast he typed up a list of names and addresses of their contacts in Springfield, Massachusetts, who were suspected of being fascists. The list included doctors, fruit vendors, mechanics, a notary public, a watchmaker, and a municipal court interpreter.

Meanwhile, he and Ullman vetted applicants for spy positions with journalist cover jobs in Lisbon, Madrid, Stockholm, and Dublin. He recommended one who had worked for him "for three months as editor of a projected magazine, since abandoned." They wrote up classified notes of interviews with candidates for German-speaking Jewish OSS positions, including a reference from Time, Inc., in Radio Center. As Downes noted later, "Some valuable results came from my counter-espionage snooping, but far more satisfactory was the recruiting of personnel for OSS staff and for overseas missions."

Against the stereotype of spies as young loners and martial-arts experts, records show that OSS recruiters valued the business skills of a mid-level manager. Interviewers considered questions such as, "Has he a good head for business? How much does he know about the transportation of supplies?" Age was not a disqualifier; the OSS wanted a workforce "diversified in respect to age, sex, social status, temperament, major sentiments, and specific skills, but uniform in respect to a high degree of intellectual and emotional flexibility." Other criteria included "absence of annoying traits" and ability to tolerate physical danger ("gunfire, bombing") and discomfort.

OSS training typically included watching a forty-two-minute orientation film in which actors demonstrated unlikely and often wrong-headed judgment in scenarios where recruits might find themselves. In one scenario, a cocky young adventurer named Charlie gets hired despite his overconfidence and laziness memorizing his cover story. The film returns to him as he gets a sobering wake-up call on the required homework and attention to detail. Charlie bides his time as a fisherman in the port of Porto, establishing his identity. When he gets his chance to make contact with Nazi operatives on the waterfront, Charlie's sloppiness betrays him, and he winds up hiding for his life in the shadows beneath the dock, isolated and alone.

* * *

That summer of 1942 Americans were rattled when they discovered that the spying ran in both directions. Eight Nazi agents arrived on American beaches, ferried there by submarines. There were two teams of saboteurs, deposited on Long Island and Florida's Atlantic coast, with detailed plans to sabotage key infrastructure and spread panic. The German agents arrived with enough explosives and cash for two years of mayhem. Their operation was dubbed Pastorius (named for a founder of an early German settlement, Germantown, later absorbed by Philadelphia). The four-man team that landed on Long Island had two immediate targets: blow up the Hell Gate Bridge and cut off New York City's water supply, threatening millions.

The saboteurs managed to get into Manhattan and were on their way to their first targets when they were arrested. One of the plotters had given the plans over to the FBI, hoping to get asylum in exchange.

When news of the Nazi saboteurs broke in early July, Americans' alarm skyrocketed. Roosevelt ordered a military tribunal, the first time a president had done so since the trial of Lincoln's assassination conspirators nearly a century before. Six of the eight spies were executed by electric chair in Washington that summer; two were sentenced to long prison terms.

Along with their primary targets in transportation and water supply, the German saboteurs had plans to hit key industrial sites, including aluminum plants near Philadelphia. With several industrial fires and accidents of the previous two years still unsolved—including the Baltimore blaze at the Crown Cork and Seal factory and the Hercules Powder factory—the Pastorius incident showed that fears about industry and national security were grounded in reality.

* * *

One fall day in 1942, Robert Ullman walked through downtown Manhattan to a lunch meeting with a bottle-cap man named Herman Ginsburg. By then, Crown Cork and Seal had joined a government committee for managing US cork resources, had been investigated for antitrust law violation of its market dominance, and was being tapped by spymasters to lend cover to American spies. An awkward position for a business—one that came with a hint of implied threat. The OSS identified Crown as a good prospect for at least two agents on the Iberian Peninsula. ("Their Spanish and Portuguese subsidiaries offer a number of possibilities to O.S.S.," Ullman wrote.) One agent would be in Spain, working as a chemist at Crown's operation north of Barcelona on the coast; another in Portugal would serve as a buyer for Crown's Lisbon branch.

Melchor Marsa may not have known that his name was being spoken as a potential espionage candidate. Over their lunch that day in early October, Ullman the OSS agent and Herman Ginsburg discussed the situation in Europe and Crown Cork's work there. In a secret memo dated October 2, Ullman assessed Ginsburg himself as a spy candidate, calling him "able and, I should judge from one long interview and Mr. Arthur Faubel's (Director of the Cork Institute) recommendations, a very trustworthy man. I told him of our need for agents in the Iberian peninsula, and he immediately entered into planning the use of his business enterprises as a cover. Mr. Ginsberg [*sic*] himself might make a very good man. A trip for some length would be reasonable. I think he would make an able agent despite his lack of any language other than English."

For the cover position in Spain, Ullman and Ginsburg discussed Marsa and his son, Melchor Jr. The OSS recruiter explored Marsa's political beliefs, later describing him as coming from a Catalan family, with American citizenship, "elderly, formerly mildly and confusedly pro-Franco though always anti-clerical."

Ullman's secret OSS memo said that Crown Cork's Portugal plant could use several people, whom the intelligence agency could help select without needing to fund them. He listed a few possible downsides: the need to involve other Crown Cork executives, their fear that involvement could hurt the company's postwar exports, and the cost of $100,000 for setting up equipment. But the memo concluded with a recommendation for two cover positions with Crown Cork.

Ginsburg himself was indeed well suited to be an agent. He was used to business travel, crisscrossing the Atlantic several times a year. He was familiar with the *Majestic*, the *Champlain*, the *Caronia*, the *Washington* and the *Normandie*, the *Aquitania* and the *Mauretania*. He liked first-class luxury liners and could chat up fellow passengers, sometimes over a chessboard. A Crown coworker later described him as "a self-made man who, by virtue of persistence, brains, and hard work clawed his way to the top. Small in stature and unprepossessing in appearance, a Jew in a gentile company at a time when anti-Semitic prejudice in corporate America was still common, he had advanced up the corporate ladder because he was needed rather than liked, and respected for his achievements even as he was resented for his brains."

In 1942 Ginsburg and his wife, Bobbie, were living in Morris County, New Jersey, in a house with her parents. Herman was commuting into Jersey City and his office at Exchange Place. He might have first met the OSS recruiter on a ship returning from Europe or at one of the Manhattan hotels he favored, especially the Plaza. As a Jewish émigré from Europe himself, Ginsburg was keenly aware of the refugee crisis in Europe and listened intently to the man from the government, for clues.

* * *

Ginsburg presented a carefully polished version of himself to the world: he was always poised and impeccably dressed in coat and tie. Through brilliance and meticulousness, Ginsburg had risen dramatically in WASP-y corporate America.

In the 1920s and '30s, Ginsburg had also led a very different life than what many of his friends and family knew. He had been married once before, a marriage that left a long and secret trail. In 1925 Herman married a young woman named Lillian Gerber, who came from a Washington, DC, family that prospered in real estate. Lillian and Herman moved into an apartment in Brooklyn—they shared a love of travel and the hope for a better world. When Crown Cork sent Herman on an extended trip to its subsidiaries in Latin America and Europe to review their accounting practices, Lillian went along. In 1928 they spent months in England, Spain, France, Belgium, and Germany.

Then the Great Depression hit, and they saw millions cast into poverty. The couple's travels for Herman's work showed them the depths of the economic devastation in Europe. Lillian became a political activist. Like many young Americans disillusioned by capitalism after the crash, she was drawn to communism as a solution to the desperate inequality she saw all around her. The couple spent a week in the Soviet Union in 1932. Herman was not convinced. They argued.

One night in 1933 Lillian, furious, tore up her husband's papers and threw away his passport. Within a year they had divorced. She remarried a man named Herbert Benjamin, who was a prominent activist and a leader in the American Communist Party. Lillian stayed active in Communist Party programs into the 1940s.

Herman, too, remarried. With Bobbie Shapiro, he made a new life in New Jersey. His international career continued to rise. Public perceptions of communism had hardened—from the idealistic alternative of the early 1930s, it was now a crueler nemesis of democracy. By 1942, when the OSS agent contacted Ginsburg, communists were considered a grave threat to America's security. With a government agent paying him a visit, Herman must have wondered if his previous life had come to haunt him.

Warily he sat down with Ullman and listened to the man talk about the threats to the American way of life. Finally, when it was clear that the focus was Europe and their meeting did not involve Ginsburg's past, he could breathe easily. He always tracked carefully the gestures of people he spoke with. He noted who brought up a topic and when. And he would run all this through his mind again after the OSS man left.

* * *

Five days after Ginsburg's first meeting with Ullman, the OSS agent reported on their second conversation. There they crystallized a plan for two undercover positions: a chemist at Crown Cork's San Feliu plant in Spain and a cork buyer for the Lisbon subsidiary. The plan was moving forward subject to approval by McManus. For the Lisbon position, they needed someone who spoke Portuguese and had general business experience. The cover salary would be around $4,000 per year. They already had their man in Melchor Marsa.

The Marsas and Ginsburgs often got together outside of work. When Marsa's son was getting ready to join the army, his parents prepared a surprise farewell dinner for him, to which they invited Herman and Bobbie Ginsburg. It was a pleasant spring evening. The two families waited for Melchor Jr.'s arrival from Baltimore at the Marsas' home in Brooklyn. The young man was driving up after several months working at Crown Cork's headquarters. After his stint in the army, he would be set to join the company in a better position. The guests waited and chatted. Seven o'clock passed, and then eight o'clock. Dinner was getting cold, so Pilar asked the guests to sit down to eat. Surely Melchor Jr. would be along soon.

When a phone call finally came after 8:00 p.m., they learned that their son had been in a terrible car accident on the highway. He was in a hospital in New Jersey, and the doctors didn't know if he would survive. Herman rushed with the parents to the hospital near Philadelphia, while Bobbie Ginsburg stayed home with the two girls.

* * *

The Brooklyn that Gloria returned to at age seventeen was different than before—in some ways smaller, and with new practices brought by the war. People were required to wear an ID bracelet with an address and telephone number, in case they were wounded in an air attack, or "if a piece of their body were found somewhere." Wartime rationing had changed shopping patterns, not only for expected items like meat but also for leather goods, which were now subject to a 15 percent tax. Shoes were rationed. Her father bought several pairs of shoes in expectation of more severe shortages. "Just in case," he told Gloria.

He was under pressure to return to Portugal for work. Crown Cork had lost access to other ports, and McManus needed Marsa to go back to manage shipping of another harvest. The *Brooklyn Eagle* reported Marsa among passengers flying to Lisbon on Pan American Airways' American Clipper from LaGuardia Field. His fellow passengers on that rare commercial flight included the director of Quaker relief activities for Europe on his way to Marseille (which was still free of the Vichy government), a British antiques dealer, a Danish ship captain, and the daughter of a British noble. The newspaper mistook Marsa for a foreigner, a "Portuguese manufacturer of cork products, returning to Lisbon after a visit with relatives in the United States."

Now Gloria was afraid for her father *and* for her brother, who spent months recovering at home from his injuries received in the car crash.

* * *

In the months after Ginsburg's meetings with the OSS recruiter, Crown Cork's defense contracts scaled up further. The McManus Cork Project boosted the company's national profile and support at both federal and state levels. That winter for the Cork Project, foresters in Morocco gathered acorns in the Atlas Mountains for the charter flight to Maryland.

Morocco, newly liberated from German occupiers by the British-American campaign known as Operation Torch, received other marquee visitors that winter as well. On January 11, 1943, a Pan American Clipper left Miami for Casablanca, bringing in utmost secrecy the US president—the first overseas flight by any American president. In heavy winds, Roosevelt's plane sailed low over the ocean. On the same day, Churchill arrived from Britain. The two were meeting for the Casablanca Conference on how to win the war. The choice to meet in North Africa was bold: it highlighted the importance of the Battle for North Africa and signaled the Allies' confidence, effectively telling the world, *We've pushed the enemy back*. One item on the agenda was the intractable problem of meeting production quotas for aircraft essential for the next phase of the fight.

The next morning in Baltimore, Frank DiCara, age sixteen, walked to the factory making B-26 wings for his first day on the job.

Roosevelt landed in Casablanca on an airfield still scarred from battle. A motorcade whisked him to the city's outskirts. The hotel, surrounded by barbed wire and equipped with a bomb shelter in the pool, was called Dar Es Saada, Arabic for "House of Happiness." FDR surveyed the elaborate furnishings and gave a whistle. Outside, Secret Service agents burned all trash to avoid leaving any clues behind. That night Churchill knocked on Roosevelt's door, informally starting their review of war plans and US industrial production. Their discussion of bombers included the new B-26 Marauder, which would be crucial to an attack on Hitler's forces. They disagreed on some points: Roosevelt expressed American commanders' reluctance to send bombing fleets for night raids; Churchill said that the British had no such reservations. But coming out of Casablanca, the two leaders agreed on a plan they called the Combined Bomber Offensive.

"This meeting is called the 'unconditional surrender' meeting," Roosevelt told reporters when they concluded on January 24. Although it was not clear at the time that the Allied advance would hold, Casablanca marked a turning point.

A month later, farther east on the North African coast, Allied forces landed in Tunisia as Operation Torch continued. They met resistance in the hills east of Tunis from German fighters under Erwin Rommel. The Allies were pushed back to the railway town of Sedjenane, nestled in cork forests on the road to Mateur. The low, bosky hills were essential to holding the crossroads of Bizerta. The cork forests of Tunisia had become a focal point of the North African campaign. On February 26, the Germans launched an offensive to outflank Allied troops at Sedjenane. Their advance was blunted by a British infantry battalion and commandos supported by the Royal Artillery, coming up switchbacks through the forests on bitter cold nights. Rick Atkinson noted:

> At dawn on the twenty-eighth, the brigade was told to seize a crossroads ten miles west of Mateur by sunset; the order meant covering twenty-six miles on Route 7 at speeds unprecedented... a road barely as wide as a single lorry and tortured with hairpin turns. The cork

harvest had left black scars around the tree trunks; stacks of curing bark awaited transport to market. Women in magenta robes thrashed their laundry in trickling creeks. . . . Squalls blew in from the Mediterranean, thickening a mud the Argylls likened to "a mixture of putty and glue."

At dawn on March 2, the British suffered in bloody fighting and fell back. That afternoon the Germans advanced toward Sedjenane, past the Fifth Battalion Sherwood Foresters. It looked like the Axis forces would seal up North Africa and keep a pincer hold on the whole Mediterranean and Europe.

For weeks the war seemed to hang in the balance in the air of Sedjenane's forest.

Then, on April 1, the Allies recaptured the railway town. US forces arrived on April 12 to reinforce the Allied positions. Armstrong Cork's Algerian subsidiary got caught in the fighting, billeting troops in several of its plants. German air strikes flattened Armstrong's Djidjelli factory. But the Allies held Sedjenane that day and through the rest of the North African campaign.

* * *

That spring Herman Ginsburg tended the roses in his garden in rural Mount Freedom, New Jersey, commuted into Jersey City, and continued conversations with his OSS contact. Crown Cork's management considered the proposal for Melchor Marsa to help the intelligence agency while working for Crown Cork in Portugal, and for Marsa's son to help in Spain at San Feliu. With young Marsa still recovering from the car accident, however, McManus demurred. In time he gave the okay only for the elder Marsa to offer his services.

In Brooklyn, with Melchor Jr. recovering slowly at home, the Marsas lived on pins and needles. In a flush of patriotism and anguish, Melchor Sr. registered with the US military just as he had back in 1917 for the Great War.

In June the OSS made an appointment to interview the elder Marsa for the position in Lisbon. An agent would assess the older man's knowledge, history, his instincts and flexibility for the assignment. It was possible, after all, that Marsa was already too well-known by the consulates in Portugal and Spain to be convincing. Yet OSS recruitment policy suggested he had the right stuff: ability to motivate others and engage cooperation and respect; administrative ability and the skills to manage an office and get along with all kinds of people. The OSS recruitment manual cites "a successful technical administrator [who] described his attitude as follows: 'I knew that my background was useful and I did not see how I could *not* do it. I'm glad to be doing it. I have felt for some time that I ought to be doing something.' There was nothing spectacular about the quality of this man's motivation... he even stipulated the duration of his assignment." Yet even with these boundaries on his commitment, he was an effective operative.

The OSS also approached several other cork industry workers that year. In August 1943, OSS agents recruited a young Armstrong Cork employee, David Sanderson. He received training that summer before shipping out to serve in "one of the most dangerous" parts of Spain—an area plagued "by considerable Gestapo activity," according to his recruiter. Diplomatic cover in the region was weak, and previous agents had been forced to flee quickly. Another Armstrong employee based in Seville, Clair Nathaniel Nell, also appears in OSS correspondence.

Marsa prepared to return to Crown Cork in Lisbon and resume work at the office on Rua Sapateiros. It was a nice historical touch, a nod to cork's long history, that Crown Cork's Lisbon subsidiary was located on Shoemaker Street. He had traveled to California to help the McManus Cork Project in late 1942. He had tromped with Crown Cork's George Greenan and with forester Woody Metcalf to inspect the older cork trees there, and he pronounced them good quality. Now it was time to return to cork's native range.

The OSS contacted Charles McManus at his home in Spring Lake, New Jersey, to confirm the plan for Marsa's deployment. Probably not even McManus, an unintentional architect of this web of commercial and national interests, saw the whole mosaic that cork had fitted together. Invention and enterprise had led Crown Cork into worldwide holdings and, through McManus's conversations with Glenn L. Martin, into aircraft construction. McManus's sons crossed Baltimore daily to check on their federally contracted bomber wing operation on O'Donnell Street. In the year leading up to the country's entry in the war, the value of Crown Cork and its subsidiaries rose 36 percent, according to its annual report, even as the company had to discontinue its commercial production of tin cans for beer and other products, by government decree.

Now spies were pressuring the crews of cargo ships coming into Baltimore and New York, forcing sailors to inform on each other. Besides the war factories pulling Crown Cork machinery into national defense manufacturing, cork extended more intimate entanglements. Marsa Sr. had coaxed his son into the cork business and introduced him to the labyrinth of managing a business across the cultures of Spain and Portugal. He had suggested to his son that working for the company in Spain could get him excused from military service and could be of even greater use to his country. But the son had insisted that he did not want to be excused from a burden that his peers were taking up. The car accident had nearly cost him his life, but it had also spared him from either dangerous choice for the time being.

* * *

Merchant seamen were paying the price in the Battle of the Atlantic. Germany stepped up its U-boat patrols, which sank more and more ships. The SS *Molly Pitcher*, one among thousands of "liberty ships" that America built during the war to replace British transports sunk by the U-boats, left Baltimore's Bethlehem shipyard in March 1943. It set out for Casablanca in convoy with a cargo of sugar, coffee, explosives, and tractors. On March 17, five hundred miles west of Lisbon, the crew heard an explosion. They had been struck by a torpedo.

The damaged ship veered into the path of others in the convoy. The ship's master rushed to the wheel room and found the helm deserted. Amid an ominous roar, he gave the order to abandon ship. In the confusion, three life rafts dropped into the water, men jumped overboard, and engines kept running. Some crew members were left behind. The ship spun in circles until the third mate, still onboard, wrangled the wheel and managed to steer clear of sailors bobbing in the waves. He tried to rejoin the convoy, but the *Molly Pitcher* was sinking too quickly.

The last crew members aboard tumbled into two improvised rafts in the pitch dark ocean. Four didn't make it. The survivors landed at Casablanca three days later.

The Germans sank a dozen cargo ships off Gibraltar and the Algerian coast in those months. The danger meant that fewer ships would be coming through the Mediterranean. The only cargo exchange between North America and Europe would be through Lisbon.

Chapter Six

# AMONG THE SPIES IN LISBON

· Marsa: 1943–1945 ·

> [Espionage is a] strange netherworld of refugees, radicals and traitors. There is neither room for gentility nor protocol in this work. Utter ruthlessness can only be fought with utter ruthlessness; honor, honesty, carefulness and sincerity must be left to the fighting forces and the diplomats.
>
> DONALD DOWNES, internal memo to OSS director Bill Donovan

By the spring of 1943, Melchor Marsa had returned to Lisbon repeatedly in the first years of America's involvement in the war. The city was awash with spies from both sides. Ian Fleming, Graham Greene, and the notorious Kim Philby were busy sleuthing for the British in Lisbon. Greene managed the Portugal desk, working for Philby, who ran the Iberian Department of MI6. Greene didn't speak the language or understand the Portuguese people, but he worked on creating a list of names of all the Axis spies in Portugal, combing index cards and sorting out bad records. His main job was to file intelligence reports coming into Britain from Lisbon, sometimes adding marginal notes like "Poor old 24000, our Man in Lisbon, charging around like a bull in a china shop, opening up vast vistas of the obvious." Greene also got involved in efforts to flip German agents in Portugal to the Allies' side. He later wrote *Our Man in Havana* based on what he learned in Portugal.

American readers of the *Saturday Evening Post* in 1943 knew that "everyone in Lisbon make espionage," in the words ascribed to one Lisboan. Americans generally knew little more about Portugal beyond the fact that it, like Switzerland, was officially neutral.

Marsa knew the Portuguese people and language much better than most outsiders. He had seen the fear in Europe, and he knew the dangers posed by German influence to sway Portuguese businesses firsthand. He had seen how the Nazis twisted arms to increase Portugal's trade to Germany, which had mushroomed from under 2 percent in 1940 to nearly 25 percent in 1942.

When he was in New York that spring, Marsa met with the OSS recruiter in New York.

Marsa was feeling keenly the passage of time and its effect. In January his eldest child, Mercedes, was married, and the wedding brought home his and Pilar's aging with fresh poignancy. Had more than three decades really passed since their wedding day? He was still rattled by the scare of his son's car accident and its effects on his son's nervous system. Gloria was fast becoming an adult, and he had great hopes for her. Marsa's feelings of duty were bound up with them and the world they were coming into.

It was a simmering June day in New York. The man introduced himself as Van Halsey, and their conversation had the antiseptic air of a job interview. They covered Marsa's childhood in Catalonia, his citizenship, and his employment record. They talked about politics and business in Spain and Portugal. Marsa expressed his eagerness to help the Allies in any way he could.

Halsey came away impressed by Marsa, his intelligence and his conversation, saying he had "quick responsiveness and an easy manner." Halsey saw a person of value.

The agent introduced Marsa to his OSS supervisor, Lloyd Hyde, in a classified memo to Washington, as a man who "has been employed by the Crown Cork & Seal Company since 1934 as the manager of their properties in Spain and Portugal. He returned to this country in 1941 and is presently awaiting transportation to return on company business." The OSS man described Marsa as being fifty-nine years old, "rugged," and in good health. Marsa, he wrote, "certainly knows his way around in business and political circles in Spain and Portugal."

Beneath that veneer, Marsa was not as rugged as he appeared. A bout with throat cancer three years before had robbed him of his voice for a time, and he lost forty pounds. He had found a cancer specialist in Spain, and the radiation treatment was successful, but it left his lungs compromised. For years he had put on a good front to cover his difficulty breathing.

The OSS memo continued, "I believe that he could be of some use to our people out there and particularly Gregory Thomas," the OSS chief of station for Spain and Portugal. The memo provided Marsa's address for contact in Lisbon: c/o Socio Gerente, Productos Corticeiros Portugueses Ltda., Rua Sapateiros 15–2nd.

From years of seeing the Portuguese police close at hand, Marsa probably knew the risks to foreign agents in Lisbon. A Canadian intelligence agent had been shot dead in the Tivoli Hotel on Avenida de Liberdade, and several agents of the German Abwehr had died in Lisbon Harbor, likely killed by British agents.

Crown Cork was well positioned for corporate cover. In addition to its natural interest in export channels, the company office in the heart of the city was just blocks from the port, and its factory in Setúbal, across the Tagus, was surrounded by port facilities and warehouses. Marsa's work gave him access and legitimate reasons to visit these crucial places. From these observation points, he could provide valuable information on trade flows and relationships that tilted toward the Axis.

OSS goals for its operatives in Spain and Portugal were threefold: (1) secure all information—military, political, and economic—about a possible Nazi coup or invasion; (2) gain intelligence about Spanish aid to the enemy; and (3) gather all kinds of background material that could be vital if the Iberian Peninsula itself became a battleground. Given the precarious state of the war in North Africa that spring, Spain and Portugal could quickly become pivotal.

Marsa prepared for an extended absence from home. Gloria watched him with concern as he packed. Her father always approached tasks with the utmost seriousness. "He wanted to do things perfectly or not at all," she said later. "That was the one thing we were taught from early childhood: You either tackle something and do it well or you don't do it at all."

Marsa gathered his files, discussed contacts with Herman Ginsburg over a meal, and made sure that Pilar had what she needed for the months ahead.

* * *

Spies from many nations trained their sights on Portuguese trade. US agencies jockeyed to read the tea leaves in trade news coming out of Lisbon. Every week J. Edgar Hoover continued to send the OSS director notes about Portuguese ship manifests and tungsten exports. Hoover was flexing his importance and sifting for clues to Portugal's leanings and how its commerce could affect the war. The British Press Office in Lisbon summarized the German-language press with clippings about reduced shipping traffic in the harbor and the growth of the Portuguese economy. It outlined in detail exports of cork, a top commodity. Cork exports soared from $300 million in 1941 (exceeded only by tungsten and tin) to $400 million the following year (110,000 tons).

The tinderbox of social and political tensions in Portugal caused added concern. One British report summarizing German press coverage noted: "The social problems in Portugal are serious.... While firms in the export trade ... are making large profits, the municipal employees and workmen are scarcely able to maintain their standard of living, which even so has always been poor."

Hitler had Nazi agents throughout the city, monitoring official and unofficial Portugal. At the Palacio hotel, where Marsa occasionally dined, four German agents routinely observed the lobby from all corners of the room with little regard for secrecy. One carried a raincoat even on sunny days. From time to time the four huddled in the middle to talk, then quickly resumed their positions in the corners, one hotel guest observed.

The Nazis were leery of a popular uprising against Salazar that could push Portugal toward the Allies. So the German press tried to spin Portugal's woes in the other direction, saying, "The English incited Portuguese workers to strike" and forced government concessions. An article in April 1943 noted shortages of essential goods and rumors that "discussions were taking place with a new American economic delegation (the first one doubtlessly perished in the Clipper accident) regarding the supply of the American forces in North Africa with food from the Portuguese colonies—food the absence of which is badly felt here." German reports disparaged US espionage in the Canary Islands off the coast of Morocco and alleged that the Americans were enlisting women as spies in hotels and restaurants and on ships and trains. German articles even suggested that the Americans had blocked Portuguese cork exports to depress the Portuguese economy; an item in the German *HF* claimed that the country's cork output "in 1943 was lower than in any previous year since 1932."

In fact, Allied spies *were* doing everything they could to dampen exports to Germany. Lisbon was the only place they could do so. As one secret OSS communiqué noted, "Portugal is the only neutral European country whose trade is not German-controlled." American spies in Lisbon tracked cork dealers who sold to both sides.

Lisbon also had plenty of double agents, intelligence operatives playing both sides. One of the more colorful was Juan Pujol Garcia, known in spy circles as Garbo. Originally from Barcelona, Garcia settled in Lisbon during the civil war in Spain, when his wife took a job with her family's company. Bored and nursing a hatred of Germany for arming Franco, Garcia, with few skills to offer the Allies, resolved to set himself up as a fake spy for the Nazis, feeding them poor information. He bought news reports at a tobacconist newsstand on the west side of Rossio plaza, tweaked the information he found there, and offered it to the Germans. After several rounds of this he inadvertently stumbled on an actual intelligence story. With few details, he sketched a British operation sending ships to help evacuate Britons from Majorca during the Nazi blockade. When his account almost aligned with an actual British-supported evacuation, Germany's estimation of his value rose. British officials scoured Lisbon for the operative who seemed to have Berlin's ear. When MI5 found Garcia, they orchestrated his intelligence feed to the Nazis to capitalize on his misdirection. Garbo gave the Germans vague tips, managing to salvage his credibility with an update that provided the correct landing details or evacuation site at a point when it was too late to hurt the Allies. As Garbo, Garcia sometimes had a suite at the Hotel Avenida Palace next to the Rossio train station. The Avenida Palace and the Hotel Aviz (where Marsa hosted holiday parties for his staff) were two of the city's premier spy hotels. The Avenida Palace reportedly had a secret passage on its top floor that led directly to the train station's platform, bypassing the main entrance and allowing Garbo and associates to arrive and leave Lisbon virtually undetected by city authorities.

The Allies and the Axis were vying for essential products through black markets as well as legal channels, and it was getting bloody. When Marsa arrived that June, activity in Lisbon's port was escalating beyond smuggling to sabotage: four stevedores were arrested for attempting to place bombs on British ships. That month too, a Pan American Clipper, the commercial plane that Marsa had taken months before, was shot down by the Luftwaffe right after takeoff from a Lisbon runway. All aboard were killed, including the British actor Leslie Howard. German agents reportedly had ordered the attack, thinking that

Winston Churchill was on board. They were nearly right. Churchill was in Lisbon and had just changed his flight plans, rescheduling to leave the next day.

Marsa found Lisbon in a grim mood that summer. Families despaired as food prices soared, and protests erupted. In July the government responded to worker unrest by putting labor issues under control of the Ministry of War. The Salazar government, leery of popular sentiment shifting toward the Allies, declared it had the power to put strikers "into closely supervised work battalions." Censorship grew even more ham-fisted: one night in late July, at the exact moment when the BBC was broadcasting Mussolini's fall and arrest in Italy, the power went out across the entire city. The utility was often strained on summer nights, but many said the blackout that night was deliberate.

Portugal's business climate was desperate, and stocks were sliding. The OSS cabled Washington: "Salazar's position is said to be becoming more critical." And again in another communiqué, "Portugal might even enter the war, after replacing Salazar. Unrest persists."

Amid the war and heightened police presence, Marsa reacquainted himself with his Lisbon staff and suppliers. He was known and respected in the business community, so he could ask probing questions to find out who suppliers dealt with. What he found was that many underlying trade arrangements had shifted. Portuguese exporters were hedging between American and German customers, uncertain which way the wind was blowing. Cork harvests and shipments continued through the tense summer as one of Europe's last neutral countries teetered between the Allied and Axis Powers.

The ride to the office took him down Avenida de Liberdade, then it was a ten-minute walk past the train station and the grand oval plaza at Rossio. The tables of the sidewalk cafés would be filled with foreigners—so many more refugees—as well as *Lisboetas*, who sat regarding the foreigners with a mixture of bemusement and wonder. Tailored suits, shorter hemlines—the fashions from across Europe—inspired local imitation and improvisations. Sometimes a local pointed out a particular jacket on the street for a tailor to replicate. Often, the glimpses at the refugees masked disdain.

On those days, Marsa trod the cobblestones at Rossio, past where Gestapo agents had seized an exiled journalist off the street, shoved him into a car and kidnapped him. The Portuguese police had done nothing. Marsa knew it was unlikely they'd try that with a foreign businessman who was simply tracking commercial activity. But as 1943 wore on, the entire city was growing more unpredictable.

At the end of the plaza, traffic funneled onto narrow Rua dos Sapateiros toward the river, with a slice of Praca do Comercio visible beyond.

A ferry ride across the wide river lay the port district of Setúbal, where Marsa had supervised the building of Crown's plant near the old glass factory in the industrial neighborhood of Seixal. Seixal's history stretched back to the Roman empire—its dockyards had built the first ships for the transatlantic trade in the late 1400s. (Vasco da Gama and his brother built their ships there.) The factory had good access to docks, warehouses, and the water, and rail access to the country's cork-producing south—a perfect observation post for wartime trade.

Marsa would meet his OSS contact at a restaurant, maybe a fifteen-minute walk from the downtown office. Sometimes it might be on a narrow side street in Alfama, a place that served good food. Melchor arrived prepared. Rarely did he see Gregory Thomas, the senior OSS officer to whom he'd first been referred. More often it was the youngish man who ran the Lisbon OSS desk at the embassy, Gardner MacPherson, who might meet Marsa at a church. MacPherson had studied at MIT and looked like the investment banker he had been before the war. Mostly, the OSS supervisors were preoccupied with the task of shuttling agents through their network of safe houses into France or to the Free French operation in North Africa, and pressing their agents for counterespionage, or X-2. MacPherson's hobby was photographing cathedrals across Europe. Lisbon's churches offered spectacular light.

Since Marsa had built up the capacity of Corticeiros Portugueses two years before, it had proved vital now that Lisbon was the only port where Crown Cork and Seal still could export to America. Marsa had supervised the new plant's construction in Setúbal, and its state-of-the-art facility had been a boon for the Portuguese industry. Then the government made a more significant investment in the cork sector and cork industry science under the leadership of a forest engineer named Joaqium Vieira Natividade.

Natividade saw that the war posed "big dangers" to cork forests as well as an opportunity for Portugal to grow its cork industry while Spain's was stalled. He worked with a laboratory headed by his colleague Almeida Garrett to bring new industrial processes from America to modernize Portugal's industry. "Nowhere was the meeting of experience, energy and greed more fruitful than in the United States," wrote historian Ignacio García Pereda of the relationship.

When the family lived in Lisbon, Gloria Marsa had visited the Setúbal factory with her father, and the lines of its sawtooth roof struck her as beautiful with the glorious Lisbon light shooting through. The plant looked much more wondrous than a factory to her. She had followed her father through on his inspection visit, and she paused as he examined sheets of cork about to be "cooked." The workers cured it with tannic acid, and she could still smell the pungent odor. Her father checked "the recipe," always testing for himself what the workers told him.

She could picture him returning to the car after a factory visit in Setúbal's port, the narrow streets near the docks thick with a smell of fish that even the flower baskets hanging from the balconies could not mask, and noisy from the bars. The Germans owned bars and brothels where bar girls and prostitutes were trained to draw out sailors' knowledge about ship schedules and convoy routes. The British owned brothels of their own for the same purpose. In wartime every kind of information became a weapon, even the mundane schedules of shipping commerce.

* * *

Each day Gloria took the subway to Borough Hall for classes at the Packer Collegiate Institute on Joralemon Street in Brooklyn Heights. It was a good place for young women to get preparation for college, and she felt intellectually challenged by the professors. She still savored the taste of international life that she had gained abroad. Gloria did well in language classes and hoped to work in radio, maybe in Lisbon after the war. With her training from a young age, she might replace the horrible accents of the American announcers she'd heard there—the ones that made Europeans laugh at Americans and that embarrassed her. Gloria's accents in Spanish and Portuguese were better than anyone else's at Packer. That was a small sampling, perhaps, but high quality.

Coming home each day, she checked the mailbox for a letter from her father with the postmark from the city she loved. His letters usually brought warm sentiments but few details, written in his beautiful Old World calligraphy, which landed with so much more force than the frenzied newspaper headlines. Pilar devoured the letters first and then shared them with her daughters and Melchor Jr.

Marsa Sr.'s assessments of Portuguese politics were to the point and remained respectful to Salazar as a thinking leader and stabilizer. Under Salazar the Portuguese had built that beautiful bridge across the Tagus and had raised up the national economy, even amid the most destructive war.

Her father included very little about his work, though. Gloria pictured him in his office on Rua dos Sabateiros, looking out the second-floor window. What did he see each evening leaving work? Whom did he talk with? She imagined him some days driving into the countryside an hour or so outside the city, to the cork groves to check on the harvest.

It helped Gloria to picture him outside the city: passing the vineyards, pastures and cornfields, and white plaster homes, talking with his assistant along the drive. Having a hearty country meal—stopping at a *taverna* for a long *almoço* with wine, a salad, and fresh sardines. Then going on to see the workers methodically attacking the trees, slicing up the slabs. They brandished their saws, pikes, and awls for the cash that the hollowed-out cork shells brought to their families.

Sometimes the harvesters piled the sheets into a makeshift house, like a ceremonial shelter for an obscure ancient ritual. Cork harvest paid better than farming. Still, rural families faced hardships that left many bitter and fueled Salazar's opposition.

One time Gloria's father and his driver left from Lisbon for a ride to the Spanish border and met the two plant managers for Spain in the no-man's-land between the two countries. Early in Spain's civil war, the chaos had made the border very fluid, with few controls. Since late 1937 legal order had begun to be restored, but the area was still not fully regulated. The porous border was never an official situation, but it opened a door to irregular trade and smuggling rings that both sides exploited. Those smuggling rings were crucial for Germany's trade with Portugal. Where the salaries of customs agents are low and the government depends on import and export duties for revenue, as in Spain and Portugal at that time, the situation creates irresistible incentives for smuggling and bribery. Notes historian Douglas Wheeler, "Given the poor pay of frontier officials, bribing was both an indoor and outdoor sport."

Gloria had photos of her father with the Spanish plant managers there, seated with a thermos at a picnic table covered in a light tablecloth, dressed in business clothes. He looked distinguished in his suit and tie, his features like those of a Hollywood star in the bright sun.

Their faces show no sign of anxiety, but there are no smiles either. In their topcoats and fedoras, they brave the chill in the strong sunlight. Maybe a business disaster hung in the air, and her father was there to give counsel. With the photograph she could picture that scene again. She could hear his voice and the cadence of his baritone, and the confidence with which he guided colleagues through worst-case scenarios. What would happen to the Seville plant if Franco fell in with Hitler? What should they do if the Nazis closed off Lisbon's port? Or what if Portugal erupted in its own civil war?

The movie *Casablanca* premiered in Manhattan at Thanksgiving in 1942, and for the following year people talked about its blend of Hollywood romance and real-world intrigue. The story of Rick (played by Humphrey Bogart) and Ilsa (Ingrid Bergman) dramatized the plight of Europe's refugees, and American newspapers noted the real-life counterparts in Portugal. *Casablanca* grew from a play originally set in Lisbon, and at least one of the film's actors, Madeleine Lebeau, had fled Vichy France by way of Lisbon, using a forged visa to board a Portuguese cargo ship for South America.

* * *

The family struggled to read between the lines of his letters, but Marsa knew what secrets to keep. He might have been surprised how this covert task he had volunteered for had opened something in himself—thoughts of his parents and his brother in Barcelona. His lost family. And unexpected fears for Pilar and their children, in case the war put him in harm's way. It made him aware, too, of them living so close to the nest of spies in New York. Maybe most of all, it burnished his love for his children, for Gloria, his youngest, whose brightness constantly surprised him. Marsa knew he could handle himself in a hard situation. But this role in Lisbon forced him to regard his own blind spots directly.

One time Marsa was asked to travel to Latin America to straighten out a shipment order. At least that was how Herman Ginsburg explained it to Pilar on the phone. As the Crown Cork International manager, Ginsburg telephoned the Marsas' home in Brooklyn and told her, "Your husband is on his way to Argentina." Pilar was surprised. "Why?" she asked. Herman said there was a claim in Argentina, a big claim for a shipment that was being refused.

During the ship's crossing to Argentina, Marsa became gravely ill and didn't leave his cabin. Eventually another passenger—a young Spanish doctor—grew concerned by Marsa's absence from the dining cabin and paid a visit. Marsa seemed practically paralyzed. Yet even as he described his symptoms, the doctor saw he was smiling. Why? he asked.

"I'm thinking this could be the best way to go," Marsa replied. "This way my family won't be faced with the horror of a death and all its details. I'll be tossed overboard and that will be the end of it."

Melchor Marsa felt isolated from his family and was suffering the aftermath of throat cancer treatment and the early stages of the emphysema that would eventually kill him. He was removed from the world he knew, and his isolation was intensified by the secrecy, the limits of what he could tell them. He also probably felt, at least occasionally, insufficient at what he was being asked to do: report on longtime colleagues, inveigle himself to get business details, uncover black market channels carefully masked by professional criminals. Doing the work of a much younger man, all while working as a manager for a company that had been his competitor earlier in his career. He had established himself as his own man in the first half of his career; now he was someone else's. Miserable aboard a hellish passage to another continent, an errand boy sent by grocery clerks to collect a bill.

The business that took him to Argentina was nettlesome. Like Portugal, Argentina was officially neutral in the war, but Germany had a considerable stake in the Argentinian economy, including ownership in military and civilian construction. Buenos Aires was one of the major Latin American ports for smuggling exports to Germany, including platinum. (In January 1944, exposure of a Nazi espionage network in Argentina led the country to cut off diplomatic ties to the Axis Powers, but that action was soon reversed by a new government.) This presence of large German firms, including I. G. Farben, caused concern among the Allies. It's possible that Marsa's financial oversight of Crown Cork holdings internationally helped provide cover for discovering more about Germany's export networks there.

The doctor aboard the ship looked after Marsa and got him to Buenos Aires, and he sufficiently recovered to complete his assignment.

He flew back to the United States on a military transport through the Andes. Flying low, he watched as white peaks rose on either side, outside the windows like two arms cradling the plane. That was one image—the Andes flanking his return home—that Marsa *did* share with Gloria.

* * *

Besides cork, OSS informants in Lisbon also traced the tangled relationships in the trade of tungsten, used to harden steel in artillery and armor-piercing shells. Portugal had become Germany's main source of tungsten. German companies used transportation hubs owned by a man named Jose dos Santos Boco, who owned a crucial tungsten and ore separation plant in Gois, a few hours north of the city. The OSS found that the Banco Espirito Santo e Comercial financed Boco and, indirectly, his business with the Nazis.

The OSS in Lisbon sent a short communiqué on Virgilio Calixto Pires, a dealer who claimed to buy for English and American firms but had no legal permit. In fact, for several years he had been buying large amounts of ore in Beira for sale to the Germans. The next month, Lomelino Lira, also of Lisbon, was found secretly buying tungsten, claiming it was for sale to the Allies; yet in a sideline he was distributing pamphlets of Nazi propaganda. Tungsten was so crucial to the German military that Portuguese officials turned a blind eye to the black market; they feared that if they cut off Germany's supply, the Wehrmacht would simply seize the tungsten or that U-boats would start sinking Portuguese ships in retaliation.

One sign of German desperation for Portuguese products was that Germany was willing to pay in gold. That February, the US, British, and Soviet governments took action to isolate Germany from global markets by declaring that they would not deal in any gold that might come from the Axis Powers and their looting. The *New York Times* reported that more and more German gold was going to pay for "such supplies as cork and wolfram" from Spain and Portugal. Before, Germany could buy on credit, but the tide was turning.

Marsa could monitor only a fraction of the ships coming through the port, but those he knew, such as the *Pero de Alenquer*, were significant. Gradually he saw how German buyers were manipulating the supply lines. His contributions fueled a secret OSS report about cork factories in Portugal. The report, ascribed to a source identified only as Z., detailed a cork factory at Margueira on the south shore of the Tagus along the road to Setúbal. The company, co-owned by a Lisbon man who also worked for the Banco de Portugal, exported cork, cork disks, and metal capsules to Germany. According to the report,

> The managing partner of the firm is Mr. Horacio Coelho, nephew of Joaquim Coelho. The activity of his factory has increased because of exports to Germany. They export to the British also. The exports to Germany are in the name of Manuel Avelar and Cesar Abrunhosa, in order to please the British authorities and prevent the firm's name from being put on the black list. On the other hand Avelar and Abrunhosa export cork disks and metal capsules to Ireland in the name of Joaquim Coelho, Lda.... In November it was reported that they were applying... for a new export license.

Whenever a second firm, Barreira & Company, wanted to sell cork to Germany, it put the transaction under Avelar's name to avoid having it traced to the firm. That way the companies managed to play both sides of the war.

The details, assembled in a straightforward narrative, helped the Allies put together a clear picture and a strategy for breaking the black markets. It was not easy looking a man like Coelho in the eye when you knew he was dealing with your country's enemy. But if this Portuguese businessman, in venting his frustrations with the convoluted arrangement and its headaches, let loose some details of a pending shipment to the Nazis, then that was something. Marsa watched the man's face, listened for hesitations. For someone who prized integrity above all, this was not easy, but maybe being older helped. Melchor Marsa had dealt with so many circumstances by then. He cleared his throat, shook the man's hand, and returned from Setúbal across the river.

* * *

In the summer of 1943, some Portuguese ship crews played a dangerous role in a US undercover operation with the code name Banana, which had major repercussions. Donald Downes, the OSS agent responsible for recruiting Marsa as an informant, sent operatives by ship to Malaga in June 1943 to prepare a Spanish rebellion against Franco. The operation aimed to restore the Republican government and to gain Spain's support for the Allies. Banana's first wave of insurgents to Malaga was followed by a second group with skills in radio operation and guerrilla warfare. The British arranged for a Portuguese trawler to cross from Algiers, and Downes met with exiled Spanish Republican president Juan Negrin in London to confirm that they were aligned. "Our paths for the moment are parallel," Negrin assured him. Downes returned to Morocco to launch a shipment of radio equipment and small arms to start a revolution in Malaga and Cadiz.

At the last moment, the British ambassador in Spain got cold feet, saying he could not risk upsetting Franco after all; he retracted support for the boat. Desperate to salvage the operation, Downes forged a makeshift flotilla of rowboats, rubber rafts, and a fishing boat and set off from the beach between Algeria and the Spanish Moroccan zone. But it was too little, too late. The Malaga team had been captured by Franco's agents, and the new insurrectionists were arrested as they landed. Under torture, several named Downes.

The Malaga incident caused turmoil in Washington diplomatic circles; the OSS director denied any knowledge of the attempted coup. When pressed, he admitted OSS complicity but spun the story, insisting the rebellion was intended to *protect* Spain against an Axis attack. This humiliation of US Intelligence cemented Spain's alignment with Nazi Germany. Portugal remained neutral but through 1944 Salazar secretly allowed German assistance to Spain through shadowy channels.

By the spring of 1944, German buyers were more desperate than ever. In April, two German agents were reported buying ore and cork through Kuehne & Nagel, a German firm based in Bremen and Hamburg. Kuehne & Nagel was responsible for transporting property seized from Jews in occupied territories back to Germany. (Led by Alfred and Werner Kuehne, the company had adapted to Nazi rules quickly, forcing out their Jewish partner, Adolf Maass, and joining the Nazi Party.) The company bought from a Lisbon man named Feliu, whom Marsa knew.

Into this stew of intrigue the US Forest Service and Crown Cork sent a young American forester, Palmer Stockwell, to study cork oak growing and harvesting operations for the McManus Cork Project. Stockwell arrived in Portugal in late 1943, where he was received by Marsa and his assistant manager, Ernesto Mas, who introduced him to Portuguese harvesters and arranged for someone to translate for the young man. Stockwell was entranced by the harvest operations. Mas helped him again three months later during a visit to Spain in early 1944. Despite the deadly tensions in Portugal during those months, Stockwell does not appear to have been involved or even aware of Marsa's off-the-books activity.

When they brought the young American to visit harvesters and processors, the drive into the Alenteju east of Lisbon took them into the gently rolling savannah, or *montado*, landscape that reminded Marsa of parts of Spain. But the towns—and a few ancient, white, plaster-walled cities like Évora—exuded a more guarded presence typical of the Portuguese. In smaller towns like Azaruja, they stopped and showed Stockwell the granges where local crews did the first boiling and flattening of the cork sheets. Here, too, one caught hints of the German buyers' desperation. Their agents got out to that year's harvest groups early, and in secret, to scoop more legitimate operations. Marsa could track them by asking the local cart driver who handled transport from the harvest fields, "Take me to where you went yesterday."

Whatever the Portuguese industry may have thought of Stockwell and the effort to grow cork in America, its newsletter that winter put a positive face on it: "Metcalf and Greenan affirm that within twenty years California will be a large and rich cork-growing region. The interest of people there grows from year to year. The cork oak trees in 37 counties promise worthy *montado* systems. Of those in full production, the groves of a Sonoma County ranch, with 50-inch diameter trees, tall, crowned, showing an indicative vigor of agricultural and agrological conditions for cork oak."

* * *

Although Marsa never told his family about any secret operations, during one of his visits home, Gloria grew curious about the special treatment that he received from government officials. It was late in the war when he was visiting Brooklyn. He was preparing to return to Portugal but frustrated by a bureaucratic delay with his passport. Suddenly he announced that he needed to go to Washington. The renewal of his passport was taking too long, he said, so he was going to have to see about getting it resolved.

Gloria was in her last semester at Packer Collegiate, and she was contemplating life after graduation. The family was planning to return to Lisbon after the war's end. She wanted to apply to work with broadcasters in Europe. Perhaps she could find someone in Washington to help secure her a position?

Her father invited Gloria to come along on the trip to Washington. She quickly agreed. When the day came to take the train, she had a moment's hesitation. Her father had been away so much those last months. Would there be awkward silences? Would his frustration with the bureaucracy turn to anger at her? Her father's temper could be withering.

But together they went to Penn Station and boarded the train. It was a pleasant ride after all, and conversation came easily enough—about her studies, about business and the world, about her brother, and about life in Lisbon. In some ways, strangely, the silver-haired father and his youngest child were more alike than the rest of the family members. They both liked the challenges of problem solving.

They reached Washington's Union Station with its grand marble hall and settled into their hotel. While Gloria paid a visit to the offices of Voice of America, her father went to the State Department to pursue his passport renewal. With the war, that process had become a bureaucratic nightmare. Gloria thought he was a bit naive to imagine that the war machinery would part simply for a businessman from a bottle-cap company.

So she was shocked later when he described how his renewal application reached the top and how a famous woman in the administration, Frances Perkins, ended up resolving it. Perkins had a voice in Roosevelt's inner circle long before he became president, and she was secretary of labor throughout his presidency. As a close advisor, she had additional duties that included oversight of the Immigration Service. Since her visit to Germany in 1933, Perkins had grown more concerned by Hitler's rise, and occasionally she advised European refugees on navigating the bureaucracy even after her official role with the Immigration Service ended in 1939.

After getting the runaround at the Labor Department entrance, Marsa reached Perkins's office, saying, as he told Gloria later, "I have to get back to Portugal. It's not coming through, and I'd like to know why." He was received politely, and he handed over his documents. Perkins asked for his file. She signed the new passport and handed it to him herself. He told Gloria, "It's amazing. When you go to the top and they know the full story, you are attended to."

This did not explain for her the surprising warmth with which official Washington received Melchor Marsa. Did this foreign-sounding gentleman in a minor industry really get welcomed and assisted by the famous woman in the Roosevelt administration? When her father told this story, Gloria thought he must be teasing her. But he did not budge from its details. She always had respected her father, but now that was being tested.

The situation seemed cruel too, in the face of her own failed interviews. The people at Voice of America told her that they didn't do hiring for the foreign offices in Washington. VOA studios around the world hired locally. It would be better for her to get to Lisbon and apply at the studio there directly.

Back home, she returned to the final months of her classes. In a few days, her father left again for Lisbon.

* * *

The war's casualties grew more devastating, and the conflict was entering a new phase of destruction. Allied troops were battling their way north up the Italian Peninsula in late 1943, and from the east the Red Army was pushing German forces back in Ukraine. Off the battlefields, shipping and smuggling remained a central focus for spymasters in Washington, who sifted meaning from the arrangements of commercial infrastructure in Lisbon's port, its labyrinths with secret entrances and blind exits. The hand-drawn map showed warehouses at Vila Pereira, on Rua Pereira Henriques where it crossed railroad tracks to Braco de Prata. There was a customs office at the quay.

One OSS communiqué was devoted to the railway quay at Vila Pereira, which masked a smuggling operation. The report, complete with a sketch showing the layout and access to roads and railheads, had the precision of Marsa's patent applications. It showed that a configuration of warehouses built by Manuel B. Vivas allowed him to gather goods coming from abroad at the quay and send them out by either of two street exits, at any time of day or night, without being detected.

Other communiqués continued to detail which dealers were selling to the Nazis under the table.

Ships' crews were more jittery than ever in the wake of the failed Spanish coup. Sailors reported on their crewmates, like when the engineer on the *San Thomé* had a visitor while in Lisbon's port. The visitor, from a Portuguese maritime police boat, came aboard and had a long talk with the engineer before returning to shore. Smuggling was the suspected topic. The ship's engineer had previously been seen "entering the house of the Japanese Naval Attaché at Lisbon," an OSS informant reported. "He is very suspect."

Corporate intelligence, gathered through day-to-day observations, helped break up smuggling rings that were key to Germany's supply lines as the tide of the war turned. The information had life-and-death consequences.

In early 1944, a Swedish ship, the *Margaret Johnson*, a 5,100-ton freighter that frequented ports in New York, New Orleans, and Buenos Aires, was docked in Lisbon Harbor. Acting on a tip, authorities inspected the ship and found a time bomb. Sappers defused the explosive and three stevedores were arrested. The police made no public statement, according to OSS communiqués. The sabotage attempt and arrests were kept out of the newspapers. But the case may have haunted Marsa. Was that the ship he had taken to Argentina? Could the crew have been looking for potential targets even then?

* * *

Intelligence agencies on both sides of the war drafted plans for suasion and destruction in case Portugal fell to the enemy. If Portugal joined Germany, British spies planned to destroy all the oil installations along the Tagus, road and railway connections to Spain, and even runways at smaller airports. And German officials used carefully calibrated methods to flip the sympathies of Portuguese officials, one by one, and make them susceptible to bribery. Something as innocent as a departing diplomat's sale of a well-maintained Mercedes to a local at a fire-sale price could open the door. Then the diplomat's replacement arrives and offers to buy the car back from the local person, at a premium due to the short supply of good cars and good tires (rubber was especially scarce). Who could call that bribery? But still, such a scheme created a feeling of obligation, just as solid as an outright payment.

The OSS staff in Lisbon was shorthanded in 1944. Several local hires were suspected of disloyalty, compromising classified documents, and listening in on phone conversations. Lois Lombard, an American staff member with multilingual abilities, hired as a secretary in June 1944, reported that one senior officer was "impossible to get along with." Lombard reported that the officer "was uncooperative and had a bad effect on the morale of the whole office staff," both locals and expats. Disgruntled agents complained that official cover for intelligence work was not useful; as one noted, State Department cover did not give them scope to visit locations without official Portuguese escorts. They needed private cover positions like the cork industry jobs, which allowed freer travel in the countryside.

Labor strikes paralyzed Lisbon in 1944, until the government stepped in and arrested factory owners who had fired some striking workers. For leverage, Salazar's henchmen held the wife of one factory executive "as a hostage" to get the business owners in line and end the strike.

Marsa faced headaches from his employees too. On top of the snooping, he was trying to run a business. Many days he returned to the apartment late, alone and tired. His health was suffering, and he wanted to rejoin his family. Once again he offered his resignation to McManus. Again McManus refused to accept it. Marsa was crucial to their success, he said. McManus Jr. called the Lisbon business "a piece of luck." Decades later he spoke of Melchor Marsa with the highest respect, saying, "Through the war, that was our contact."

* * *

Portuguese politics complicated Marsa's task of keeping that supply channel open. Salazar's government stood on the knife edge of neutrality until the bitter end. On April 30, 1945, just days before Germany surrendered, Salazar was one of just three European leaders to send condolences on Hitler's death. Lisbon erupted in public protests. Salazar's opponents prepared for the Allies to sweep his dictatorship out of office.

VE Day brought the exultations and relief of the Allies' victory. Months passed. Finally, in August, the Allies signaled that they were willing to deal with the dictator. Winston Churchill accepted Salazar's invitation to visit Portugal.

That fall Pilar Marsa sailed to Lisbon over a free Atlantic to reunite with her husband, whose health had been a source of mounting concern for her. After several months in Lisbon, the couple returned at Thanksgiving on a liberty ship, the USS *Henry Hadley*. Gloria was visiting her sister and brother-in-law in Mexico at the time, so she didn't meet her parents at the dock.

Expecting to return to Lisbon again in a few months, Marsa left the company's Lisbon branch in the hands of his younger associates, including Ernesto Mas. Marsa would never speak to his family about his recruitment by US intelligence agents or what he did during the war to help them. Like a good spy, Melchor Marsa sealed off that chapter of his life and sailed for home.

This image of firefighters at Crown Cork's Highlandtown factory blaze appeared in the September 18, 1940, edition of the *Baltimore Sun*. Courtesy of the *Baltimore Sun*

Cork trees growing in Portugal (jnc 3.6-1).
Copyright Junta Nacional da Cortiça, archive José Neiva

A Portuguese official watches as workers harvest cork (jnc 4.3-38).
Copyright Junta Nacional da Cortiça, archive José Neiva

Ships loading cork in Lisbon's port (jnc 5.8-25).
Copyright Junta Nacional da Cortiça, archive José Neiva

At the 1939 New York World's Fair, the Road of Tomorrow exhibit featured a smooth ride cushioned by cork. Courtesy of Ford Motor Company

Melchor Marsa Sr. in Lisbon. Courtesy of Gloria Marsa Meckel

Gloria Marsa in 1940. Courtesy of Gloria Marsa Meckel

Algiers in the 1920s.

Melchor Marsa, Bobbie Ginsburg, and Herman Ginsburg in New York.
Courtesy of Gloria Marsa Meckel

Our Lady of Pompei procession in Highlandtown. Courtesy of Angela DiPasquale Knox

The Los Angeles plant of Western Stopper Co., a Crown Cork and Seal subsidiary, 1940s. Courtesy of Baltimore Museum of Industry

Cork stripping in California with the McManus Cork Project. Giles B. Cooke collection, Swem Library, College of William & Mary

In Lisbon, a galleon and the Sphere of the Discoveries at the Portuguese World Exhibition in 1940. Courtesy of Paulo Guedes

Workers cutting cork in Portugal. Copyright Junta Nacional da Cortiça, archive José Neiva

The cargo ship *Pero de Alenquer* unloading cork in New York harbor, August 1941.

Woodbridge Metcalf and Melchor Marsa examine cork at the Lisbon port.
Courtesy of Gloria Marsa Meckel

Melchor Marsa with his Spanish Crown Cork and Seal manager Sr. Bonal of San Feliu de Guixols, conferring near the Spanish border. Courtesy of Gloria Marsa Meckel

THE SATURDAY EVENING POST

## "Listen, soldier, it just doesn't make sense!"

BILL: But I'm telling you it's true. Take a look. Here's their trade-mark right on this shell.

JOE: Get hep to yourself! Armstrong is a linoleum outfit. Guess I ought to know. My old man's been selling Armstrong's Linoleum in his furniture store for years.

BILL: Aw, must be a different company. Armstrong Cork makes bottles and corks and caps. Used to use their stuff all the time in the drugstore where I worked.

HANK: (*Ambling up from gun*) You guys are both dizzy. Armstrong Cork is a building materials company, see? Before I got into this man's army, I was a carpenter back in Ohio, see? And I nailed up so much Armstrong's Temlok insulating board that I'll never forget that name!

BILL: All I know is that every prescription bottle in McCabe's drugstore had this same Armstrong trade-mark stamped right in the bottom.

HANK: Must be a lot of different companies with the same name.

JOE: Yeah, and none of 'em sounds like the kind of an outfit that could turn out steel shells. It still doesn't make sense to me!

WE DON'T BLAME ANYONE for being a little confused about the Armstrong Cork Company. Or for thinking that the many different products that bear the Armstrong trade-mark must be made by many different companies. Folks in the textile mills come to us for roll coverings. Meat packers know us for cold storage insulation. Car makers for gaskets and seals. Steel mills for insulating fire brick. Architects for dozens of building materials from floors to acoustical ceilings.

But all these hundreds of diversified products are made by one company—a company that today is also producing shells of many types, cartridge cases, bombs, bomb racks, projectiles, aircraft parts and assemblies, camouflage netting, and dozens of other vital products for war . . . and for victory.

Lancaster, Pa.; Camden, N. J.; Pittsburgh, Pa.; Millville, N. J.; Beaver Falls, Pa.; Fulton, N.Y.; Dunkirk, Ind.; Philadelphia, Pa.; So. Braintree, Mass.; Gloucester, N. J.; Pensacola, Fla.; Keyport, N. J.; South Gate, Cal.

An ad for Armstrong Cork in *Saturday Evening Post* plays off the industry's unexpected role in the wartime manufacturing effort. Courtesy of Armstrong World Industries

For California Arbor Day 1944, Governor Earl Warren plants a cork tree. Giles B. Cooke collection, Swem Library, College of William & Mary

Woodbridge Metcalf with the cork project truck. Giles B. Cooke collection, Swem Library, College of William & Mary

At the Maryland Arbor Day ceremony, brothers Walter and Charles McManus Jr. flank Governor Herbert O'Conor. Giles B. Cooke collection, Swem Library, College of William & Mary

Melchor Marsa (left) inspects cork tree health with the McManus Cork Project at Chico, CA, 1942. Courtesy of Gloria Marsa Meckel

W. Michael Blumenthal and his sister in 1946. Courtesy of W. Michael Blumenthal

Serviceman Frank DiCara in 1945. Courtesy of Frank DiCara

Cork trees in Flat Creek, South Carolina, with the grandchildren of Glenda Pittman Owens, whose father planted the tree in the 1940s. Courtesy of Glenda P. Owens

Chapter Seven

# FROM THE FACTORY TO THE FRONT

· DiCara: 1943–1945 ·

> "Bottle caps!" Childan had exclaimed without warning.... "We used to collect the tops from milk bottles. As kids. The round tops that gave the name of the dairy."... [P]robably it was still possible to obtain the ancient, long-forgotten tops from the days before the war.
>
> PHILIP K. DICK, *The Man in the High Castle*

Frank DiCara left his shift at the bomber wing factory and walked back through Highlandtown, past the sewing company on Grundy, where his sisters worked. His hands still vibrated with the buzz of the drill. Every few months he got a raise of five cents an hour. By early 1944 he was making eighty-five cents an hour at McManus's factory—not bad for a seventeen-year-old. It was nearly three times his starting pay there.

That year US factories produced war materials at record levels, following labor strikes the previous year by workers in the railroad, coal, and steel industries. Workers struggled between their desire to boost the war effort and their call for a fair share of the improving economy. Overall production quality improved. The B-26 bomber that DiCara helped to assemble faced a first impression among pilots as being balky and a "widowmaker." A pilot flying a B-26 had to maintain an exact airspeed to land—and that 150-mile-per-hour requirement intimidated many pilots who were accustomed to landing at much slower speeds. Later flight crews said that the planes became more stable after manufacturers made the wings a foot or two longer.

Highlandtown and the war had kept on going while Frank's brothers fought in Europe. Another Highlandtown factory, one that manufactured refrigerators, had burned down (no mention of sabotage). A new report from the ACLU ranked Baltimore among two dozen communities with the worst race relations in America, where the potential existed for riots. The report was titled "How to Prevent a Race Riot in Your Home Town." And in May the governor announced that nearly 172,000 Maryland men were serving in the armed services. Speaking to a Baltimore American Legion gathering that included representatives from Highlandtown, Governor Herbert O'Conor summoned the number to emphasize the need for the "GI Bill of Rights" then before Congress. The number was just a small indicator, he said, of "what war has meant to date and how great a shock it has been to the ordinary life of all our people."

As Christmas 1944 approached, Frank received another pay hike—to a dollar an hour. He also received in the mail an unwelcome notice from the draft board. Waiting for him when he got home from the factory, it told him to prepare to report for induction into the army.

His mother was beside herself. Her three older sons were already serving in the war in Europe. How could they call up her youngest too? How much was her family supposed to give this country? She besieged anyone who might help get Frank excused from service. His sisters sent letters to officials. Their pleas got him a short delay—probably due to the fact he was working in a war factory—but they couldn't get him exempted.

Frank was ordered to report for duty at the Fifth Regiment Armory, the vast fortress of a building up on Division Street, not far from Baltimore's Penn Station.

On a Friday just days after Christmas, Frank said his good-byes to Irma, to his sisters, and to his mother. Then he made his way across town and approached the armory building, its rows of stone turrets standing at attention. He had just turned eighteen and had never been out of Baltimore.

With other recruits, Frank boarded a train to Fort Knox, Kentucky, for basic training. At Knox he crawled across the ground, keeping down and careful not to shit his pants as a tank rolled over him. Next they put him *in* a tank, with an instructor showing him how to work the gearshift with two hands. It was a basic exercise to familiarize recruits with equipment. Frank looked out the tiny, mail-slot-sized window before him as the tank bounced over rutted land. Some of the dips were six feet deep, pitching the tank in a way that made the "windshield" useless. He couldn't see anything. He had never driven a car.

He was becoming a member of the 640th Tank Destroyer Battalion.

The recruits got a raft of vaccinations against tropical diseases. The needles struck fear into some of the young men. The tall guy in line ahead of Frank told him, "Don't worry about it. Nothing to it." But when they gave the big guy the needle, he fell flat on his face. One of the shots made Frank's arm sore for days.

They marched for miles, came back to the base, and found blisters across their feet. One night someone said, "Frank, we're having a GI party in the barracks." Frank said, "A party?"

"Yeah, we're all going." When he got there, he was handed a mop and a bucket and told to scrub the floor.

After weeks in Kentucky, he boarded another train—this time for a longer ride, to California, in boxcars. The fields they passed were blooming with spring green. With the freight car's door open, they rumbled through midwestern farmland. Frank saw farmers waving in the distance. The soldiers waved back, rolling away.

After several days they reached Fort Stoneman in Pittsburg, California, near the San Francisco Bay. Stoneman was the staging point for soldiers bound for the Pacific. He was surrounded by an astounding variety of guys from many backgrounds. Some got shipped on to Guam and the South Pacific. Frank didn't know where he was going. He waited for his orders.

* * *

Thirty miles west of Stoneman, walking across the Berkeley campus one spring morning, Woody Metcalf felt far from the war. Students were walking to class, the bells rang, and forest fire plans were drafted as usual. At the post office, Metcalf found a letter from his former student and employee Ralph Waltz, serving on the USS *Neosho*. The envelope was stamped "Passed by Naval Censor" and stippled with black marker.

Even before he read it, the letter moved Woody as he recalled how, just months before, he and Ralph had traversed the sun-dappled California forests. They had stripped cork trunks across Napa County together and scaled the steps of the Berkeley Student Union to give presentations on topics of recreation and education in conservation. Woody had edited Ralph's reports on wartime shortages and their impact on the price of laurel bay leaves from California forests. He had watched the young man grow into a professional.

Now Ralph's angled handwriting described the tedium of life aboard ship, the slow days punctuated by sudden frenzy. Ralph recalled their fire management classes, saying that he and his crewmates had to be very fire conscious themselves now. As ship's engineer, he encountered "tricky problems and situations," including a machinery casualty that kept "this forester hopping." Ralph also told Woody about his young wife's return to Ohio to be with her family when their baby arrived, and the pain of the long separations. He described the ache that he and the other men on ship felt awaiting letters from home more than anything else. "At least a letter certainly brings new life with it."

Finally, Ralph said he missed working with Metcalf in the forest, a memory prompted by a report that Metcalf had mailed him, "Cork Oak Planting in California." Reading it, the sailor "couldn't help but recall many of the interesting trips and experiences that were involved." Even statistical figures triggered memories. Did the dream journey to South America in search of cork trees ever happen? he asked. "Thought maybe Superintendent 'Corky' Greenan might have made the trip."

Metcalf was now working with Palmer Stockwell, his postdoc researcher, on measuring and weighing cork harvests from Victory Park. As an educator, Metcalf spoke with potential cork growers in Mendocino County about the benefits of planting seedlings. He also volunteered with a shelter committee, arranging beds and mattresses for temporary residents and migrants in the Bay Area. He organized a conference of farm advisors in Berkeley to discuss wartime problems like tire rationing, labor shortages, and victory gardens. To bolster home-front morale, he boosted spirits at conferences with cribbage games and songs, leading sing-along versions of "Hail to California" and "Alouette," the jaunty tune that American soldiers in World War I had learned in France. And he conferred with George Greenan, Crown Cork's West Coast manager, about gathering more cork oak acorns, aiming for four thousand pounds "if that number can be collected."

Still, sometimes Metcalf felt gnawed by doubts about his contributions to winning the war and whether the cork project was worthwhile. Simple errands required more time than before due to the wartime scarcities of gas and tires. Trips by streetcar dragged ("took a full hour and very crowded," he grumbled in his daybook), and when he drove, he had to allow extra time for traveling thirty-five miles per hour, "the new tire-saving speed. It is pretty slow on a long drive." Down the valley, Metcalf checked cork oaks discovered growing at Kingsburg and tracked down their owners.

Metcalf spent Sundays tending to home, picking vegetables in his family's victory garden, or patching a wall or corralling his kids to help "until they folded up." At the end of the day, Metcalf would sit down to scribble edits on a draft article about cork oak for the utility company's newsletter. Public education was as important as research, in his view. With articles and newsreels hammering the link between cork needed for the war and tree growing, the public was becoming aware of cork's role in national security. Local news items about stripped cork trees appeared in Alabama's *Cullman Banner*, and Armstrong Cork's full-page color ads splashed nationwide in the *Saturday Evening Post*. One Armstrong ad caught readers' attention with an illustration of bare-chested GIs standing waist-deep in jungle on a nighttime operation, carrying artillery shells to mortar emplacements, and a snippet of their dialogue:

BILL: But I'm telling you it's true. Take a look. Here's their trade-mark right on this shell.

JOE: Get hep to your self! Armstrong is a linoleum outfit. . . .

BILL: . . . Armstrong Cork makes bottles and corks and caps. Used to see their stuff all the time in the drugstore where I worked.

HANK: (Ambling up from gun) You guys are both dizzy. Armstrong Cork is a building materials company, see? Before I got into this man's army, I was a carpenter back in Ohio, see? . . . Must be a lot of different companies with the same name.

JOE: Yeah, and none of 'em sounds like the kind of an outfit that could turn out steel shells. It still doesn't make sense to me!

The ad explained that it was the same Armstrong, saying the company "is also producing shells of many types, cartridge cases, bombs, bomb racks, projectiles, aircraft parts and assemblies, camouflage netting, and dozens of other vital products for war . . . and for victory."

This was just as true of Crown Cork's production of B-26 bomber wings. Frank DiCara looked up whenever a Marauder flew overhead and would tell anyone in earshot that he had helped put it together.

When he had worked at the plant, moving down the line drilling holes into each wing, focused on placing the drill precisely and delivering the bore straight, Frank had imagined the plane was heading to Europe, to help his brothers. It would fly over and dole out "death by the bellyful," like the newsreels said, to the krauts who lay waiting for Angelo and Joe Jr. Frank's B-26 would lend air cover for Angelo coming ashore in the D-day invasion. Frank's mental pictures came from comic books and movies, but he knew it was happening.

And it was. On D-day, Stephen Ambrose wrote, "The plane that did the most damage was the B-26 Marauder," especially at Utah Beach. The improved accuracy of the Marauders' targeting paid off in their bombing raids on bridges and railyards that day. The B-26 continued to prove itself, gaining acclaim as "the chief bombardment weapon on the Western Front," according to a 1946 military dispatch. The plane ended the war with the best safety record of any bomber in the US Army Air Forces.

Many years later, Angelo would talk with Frank about D-day just once—about how lucky he felt to survive it. Their conversation left Frank with images of young men putting on their helmets, jumping out of boats, and drowning on the way to shore.

Americans on the home front channeled their desire to help win the war into manufacturing. The city of Berkeley decided to contribute a victory ship to the cause, "destined to sail the blood-stained Pacific" in the words of the *Berkeley Gazette*. Local businesses like the Sather Gate Bookstore and University Radio Shop, alongside Berkeley Public Schools and the chamber of commerce, contributed to build the supply ship in a Richmond shipyard. The keel was laid and in a feat of patriotism, the USS *City of Berkeley* was launched in January 1945, an event featuring a speech from the mayor and the high school a cappella choir performing "Smooth Sailing."

One of the employees at University Radio was a teenager named Philip Dick, for whom the owner, Herb Hollis, was a sort of father figure. Hollis was an independent small businessman, and his staff was a hardworking and creative assortment, doing radio repairs, selling appliances like phonographs, washers, and dryers. Phil Dick started working for University Radio two summers before and stayed for almost nine years, the only steady job he kept in his whole career, apart from science fiction writer. The backdrop of wartime fear seeped into the paranoid atmosphere of books he wrote later, from *Do Androids Dream of Electric Sheep?* to *The Man in the High Castle.*

At that time, Italian Americans and other "enemy aliens" in California were getting rough treatment. Earl Warren had relentlessly pursued relocation of Japanese Americans. Well over 100,000 Japanese Americans were removed from their homes and sent to internment camps. Warren drew a racial line between them and German and Italian Americans, but all three groups were scapegoated, and in the competition between the FBI and other agencies to show which was most aggressive about securing America, all three groups suffered. The "enemy alien" designation restricted the movements of over 600,000 Italian Americans nationwide; around 10,000 were detained and relocated.

Nino Guttadauro, an accountant with US citizenship, had been on an FBI watch list since September 1941, when his name appeared on a letter signed by J. Edgar Hoover that stated, "It is recommended that this individual be considered for custodial detention in the event of an actual emergency." Eleven months later, Guttadauro received a custodial detention card and ordered to leave his California home and the western states. His eviction proceeded despite a letter in his defense from the US assistant attorney general, stating "there is not sufficient evidence upon which to institute a criminal prosecution against the subject at this time." Still the FBI did not soften its stance. It ordered Guttadauro to report to an individual exclusion hearing board in San Francisco that fall. If he failed to appear, he could be fined $5,000 or sentenced to a year in jail or both.

When he showed up at the Whitcomb Hotel for the hearing the morning of September 8, Guttadauro was told that he would not learn who his accusers were, nor would he receive details of the accusations. He would not be allowed legal counsel. The suite on the hotel's fourth floor was a bizarre location for an official proceeding. It lasted less than an hour. Guttadauro's presence in California was declared a threat to public safety. Officials barred him from traveling to or living in more than half of the United States (anywhere near a coast, where he might abet invaders). The FBI pressed again to take away his US citizenship altogether, a process called "Denaturalization Proceedings." For nearly three years the investigations, interrogations, and hounding continued, as Guttadauro and his family moved from state to state looking for work. He settled in Salt Lake City, where they knew no one, for a grocery clerk job.

Guttadauro's internal exile lasted until the spring of 1944, when the exclusion order was rescinded. The ordeal left the family in tatters, financially and emotionally. Guttadauro's son Angelo later said, "We had become, by military fiat, a family of involuntary gypsies."

Even Joe DiMaggio's parents weren't spared. While their son, the Yankees' slugger, was the toast of New York, Gen. John DeWitt, a leading officer in the Western Defense Command, pressed for arresting Giuseppe DiMaggio, who had lived in the United States for forty years but never applied for citizenship papers. DeWitt wanted to make a point: "No exceptions." The FBI stopped short of arresting Giuseppe, but he and Rosalia DiMaggio, like their neighbors, had to carry their "enemy alien" photo ID booklets at all times and needed a permit to travel more than five miles from home. Giuseppe was barred from the San Francisco waterfront, where he had worked for decades, and suffered the further humiliation and hardship of having his fishing boat seized by the government. Months later, when officials let the elder DiMaggio return to the docks, the *New York Times* finally reported the episode. Keeping a light tone, the *Times* said in June 1942 that DiMaggio senior "may return to Fisherman's Wharf to keep an eye on Joe's restaurant," along with the other Italian Americans who "had been barred from that picturesque district." The short item noted that "compliance with curfew, residence and travel restrictions is still required."

Just outside Fort Stoneman, where DiCara arrived, Italian American families in Pittsburg were suffering the government's enforced relocation. Rose Viscuso was almost fourteen years old in 1944. Two years before, her whole neighborhood had been evacuated from Pittsburg to Concord, about twenty miles away, despite the fact that her father, a US citizen, worked in the defense shipyards. Months later Rose's mother put the girl on a Greyhound bus headed back to Pittsburg to see what news she could find of their old home. Might they be able to move back?

Rose traveled the miles to their old community like a stranger, but in Pittsburg she got good news: the family could return. When she told her mother and aunt, they burst into tears. "Momma sent me on to alert the others in a one-mile radius," Rose recounted later. "I can remember knocking on doors and shouting, 'You can go home now!' and the excitement of it all. . . . Paul Revere rides again!"

Executive Order 9066, which allowed the government to imprison German, Italian, and Japanese Americans without charges or trial, and seize their homes and businesses, was never successfully challenged during the war. It stood for more than three decades until President Gerald Ford rescinded it in 1976. The investigation into its effects on people's lives led to passage of the Civil Liberties Act of 1988. Still, the effect on Italian Americans remained largely unknown for decades. Many simply wanted to forget the episode.

* * *

Inside Stoneman, Frank DiCara got his shipping-out orders. He mailed a few last letters before leaving. For the rest of his life he would associate the tune of "Sentimental Journey" with the line of grunts he followed up the ramp to the ship, his duffel bag over his shoulder, lumbering toward the enemy beyond the sea. He heard the dirgelike tune played by a band there, sending them off to the war's cauldron. Even as he boarded the ship, he realized the absurd joke of the band sending them off by playing "Sentimental Journey," the bitterest irony. It was the anthem for GIs coming home from Europe, the tale of someone about to make the passage for an emotional homecoming, "like a child in wild anticipation." Yet the tune is almost a slow drag, a near monotone of dread, and fear of changes that the singer will find at home:

> Gotta take that sentimental journey
> Sentimental journey home.

As soon as he was on the ship, Frank threw up.

The captain came on the intercom and announced the blackout rules: no smoking, no lights after dark, no noise or banging. Japanese submarines were listening and could hear them with sonar. You had to keep it quiet and dark. This terrified another draftee named Thelman. "You're not the only one," Frank told him.

At night he lay awake in his bunk, listening to others sniffling in the dark: boys. Scared and not knowing if they would ever return. There was nothing said to console them; words would have been blowing smoke anyway.

The troop transport that Frank DiCara boarded in California in early 1945 steamed across the Pacific through heavy seas. It docked in

Hawaii at Pearl Harbor, where the men could see the destruction from the attack more than three years earlier as they came into port. Young Frank, the greenhorn, got a taste of shore leave. He hit the bars with a group of sailors and GIs and woke up the next day with a dollar-bill-sized tattoo on his right bicep that said "Frank."

After another forty-five days at sea, Frank reached the Philippines and joined MacArthur's legions taking back Southeast Asia from Japan. When his ship drew within sight of the shore, the danger increased. The captain delayed the troops' landing until a moonless night—pitch black, no stars out. When ordered, Frank climbed down ropes on the side of the ship, hand below hand, until his feet reached the amphibious duck bobbing beside the ship. Absolute darkness, the sound of water. He was told to put his hand on the shoulder of the soldier in front of him—the only way to stay in formation as they went ashore.

It was a night of terror for the invaders.

Retaking and occupying the islands held more eye-opening moments for a boy from Baltimore. He saw starvation in the faces of locals living outside GI camps on Leyte, Mindanao, and Tacloban. He saw families sorting through the camp's garbage and waste dumps for food. After four years of war, they had nothing. Outside the encampment, enemy mines and snipers reportedly lay in wait. Across a ravine, Frank saw five human heads on a branch, posted as a warning.

When they felt safe to leave the base, Frank and a buddy rented a hut in town and brought rice to families and a couple of girls who lived there. The families didn't have running water—household water came from a shared spigot a half mile away.

Once in a while the camp showed movies to lift morale. Frank escorted some of the local girls to the shows, including *Manila Calling*. Another movie was about winter in Manhattan. The snowflakes falling on the city streets struck everyone mute with longing.

GIs across the archipelago were bracing for Operation Downfall, the big, two-part invasion of Japan. Overhead, bombers made practice runs. Everyone in camp looked up when the flyboys came in, sometimes on near-vertical dives. On the wings Frank had helped to make. Otherwise, Frank didn't think much about the bomber plant or his former coworkers. It was already like another lifetime. He thought instead about his brothers and the news that the war in Europe was over.

Some pilots were horsing around; some might be returning from bombing missions. Were they preparing for the last push?

There were rumors that MacArthur would be promoted so he could command the whole show. It would be the largest armada ever assembled, probably attacking the south part of Kyushu.

Frank received orientations on what to expect when they attacked Japan. The tank training sessions resumed in fits and starts. The engineers were doing tests to see if Japanese soil was firm enough for tanks to move and not sink on their treads. Frank's buddies were betting the tanks would sink. Someone called them "rolling coffins."

For Frank the occupation of the Philippines held biblical-scale terrors of its own: an earthquake, bombing in Manila, malaria, and hurricane-force winds in which locals fled with babies in their arms to escape rising waters. They tied themselves in the trees, braced for the surge. Frank fell sick with malaria and took the pills issued to him. He heard he would probably get it every seven years after that.

And one day, a real seven-year locust plague struck. It was a quiet day on Mindanao. Frank was reading in his tent while outside the tropical sun beat down. When the sky suddenly turned dark, he looked up and saw the air filled with big locusts. They were tumbling out of the sky, falling into the tent. When the air eventually cleared, he looked out and saw a bare hilltop that had before been covered with wild grass.

Frank did not put any of these spectacles in his letters home to his mother, to Irma, or to the few other girls he wrote. But he kept the letters flowing, and he cherished the ones he got from Highlandtown. Irma sent him news and wrote about her job at the hospital, where she and her sister worked at the blood bank. He would always remember one postcard, stamped from the "State of Love, City of Wishes, 19 Hugs and 65 Kisses."

Frank didn't know if he would see combat. But what he saw in those days as they prepared for the push to invade Japan filled him with dread enough. Had he left poverty in Highlandtown only to die in the Pacific? His factory job should have opened a door to success, but instead he was here where he could be snuffed out any day. Meanwhile the factory and the McManus enterprise rolled on to new levels of success, growing trees and business across America.

# PART 3
# BEYOND VICTORY

Chapter Eight

# POLITICS AND GASOLINE

## · McManus: 1942–1946 ·

During World War II cork was a material of critical military importance, and the supply had to be carefully rationed. Thus an emergency supply of this material on trees growing within the limits of this continent may be of very great importance in a future emergency.

WOODBRIDGE METCALF

Once the McManus Cork Project got under way to fill a gap in national security, it gained momentum. Charles and Walter, back from his stint in the Air Force, were both now working at Crown Cork and checked in with Giles Cooke on the progress of seedling production, testing of the California cork oak varieties, distribution of acorns and seedlings to curious kids and 4-H clubs, and endless publicity. Politicians lined up to be photographed with a silent but visible patriotic effort that appealed to the citizenry and demanded so little time or argument. Everyone seemed to love trees.

The cork project demanded a fair amount of time from those managing the campaign. The McManus sons traveled to more than a dozen Arbor Day festivities and smaller ceremonies across the country and glad-handed state politicians. It was the biggest public relations campaign the bottle-cap business had ever seen. Having been a skeptic at first, Charles Jr. had come to enjoy the events and the crowds that they drew.

One of the most memorable was Louisiana's Arbor Day celebration in Baton Rouge. Walter attended with Cooke, arriving for the late January ceremony, where an official unveiled a bronze plaque. "Presented by Charles E. McManus," it said, "to encourage and promote the culture of cork oak trees" across the state. What other state boasted a Country Music Hall of Fame governor? Cooke and Walter McManus shook the hand of the Honorable Jimmie Davis (songwriter of "You Are My Sunshine") over a seedling and shovel, and the eighty-piece Baton Rouge High School band played. The crowd of hundreds filled the rolling swath of hillside before the monumental Louisiana statehouse.

"Forward-looking citizens of our country believe that we can produce at least our minimum requirements of cork in our own country," intoned the head of Louisiana State University's Forestry Department. He described how cork trees growing on farms alongside annual crops could bring new income to Louisiana farmers. "We believe the time will come when Louisiana will be a recognized producer of cork."

Governor Davis waxed poetic about the state's fifty thousand people employed in forestry: "We have a rich heritage in Louisiana's forests," he said. Yet that employment was not distributed equally, and many were shut out from those jobs. The war had seen increasing migration by black Louisianans moving north to leave Jim Crow segregation for better job opportunities.

The patriotism of the cork project gave Arbor Day a higher profile across the South. Governors in South Carolina, Georgia, Mississippi, and Alabama lifted shovels to have their photos taken next to new cork seedlings on the capitol grounds. The gesture became testimony to the self-reliance (and economic hopes) represented by the young trees. The South Carolina governor went on the radio after the planting ceremony in Columbia, emphasizing forest conservation as the state's core value. "Our state is the first to have cork trees planted in every county. We are the first to celebrate Arbor Day by the planting of a cork tree in a statewide program."

"Cork planting will continue into the postwar period," he proclaimed. "Our government will be assured sufficient cork for our Army and Navy and other essential requirements." Charles Jr. knew that the South Carolina bottling companies had a representative in the state legislature who watched out for them, and somehow the bottling industry got a mention in the governor's remarks. Students in every city school in Columbia received cork acorns to plant.

In Alabama, where the state forester distributed more than ten thousand seedlings, the governor took the stage to highlight his state's forests and their national significance.

Charles Jr. marveled at the campaign's popularity. Every ceremony involved at least several clubs, principals, and school groups, with music provided by a local high school band, usually a big one. As part of the crew taking this pageant from state to state, Charles had time to reflect on his father's and his own motives with the cork project. In a master stroke, it brought the war's economic and security reality home to Americans who knew Europe only as a place where bad things happened, and where their sons were risking their lives for freedom. Charles still felt that all the speeches smacked of a kind of self-promotional grandstanding more suited to Armstrong's Henning Prentis, as did the truck with "McManus Cork Project" painted in huge letters on the side, driving coast to coast, and a surprisingly large publicity budget. But he thought the plantings and the motivations behind them were in earnest.

By then Crown could use good publicity. The rumblings against bottle-cap producers as a cartel had coalesced in 1941 with a federal price-fixing case against the Crown Manufacturers Association of America, brought by the Federal Trade Commission. That was still working its way through the courts. And the defense contracts, begun when every company was needed for the war effort to meet quotas, were starting to look like make-a-buck ventures, and rumors of profiteering were getting stoked in the press, with frequent cartoons mocking a big corporate fat cat labeled "profiteer."

In March 1944 California joined in the cork campaign when Earl Warren, by then governor, grabbed a shovel to plant a cork seedling at the state capitol. Warren's proclamation invoked patriotism, pioneer days, and the state's contribution to modern civilization. He declared Conservation Week "to emphasize that we can better direct our full power against the enemy and at the same time build for the future by fostering [conservation] programs."

The seedling Warren planted in Sacramento, one of twenty-nine thousand raised in California's state nursery, would "stand as a symbol of the interest and cooperation of hundreds of tree planters who in the last four years have received and planted nearly 150,000 cork oaks in 40 counties of California." State and Los Angeles County foresters hauled seedlings for planting to "qualified landowners," gathered millions of acorns from trees across the state, and peeled California's majestic cork trees. "More than ten tons of cork have been stripped from California-grown cork oaks and tests have shown most of it to be of excellent quality," the ceremony noted. The vision for the future was big: a million cork oaks planted across California in ten years.

Cork was ensconced in the state's long-term environmental security plan.

* * *

Woody Metcalf had devoted much of July 1943 to the cork program, measuring the giant trees he had found, researching their growth, and raising seventy thousand seedlings from acorns in pots for transplanting. On July 23 at nine o'clock in the morning, he met with George Greenan, and with a movie crew they drove to Napa State Hospital to make a promotional film about growing and harvesting cork. When they reached the hospital grounds, with the cameras rolling, his team assembled with their pikes and saws, as he wrote in his daybook, and "we stripped the big cork oak for the movie people. The bark came off well on the upper branches but it was a hot day 93 in shade."

The stripping operation at the hospital continued for several days. Metcalf was sobered by this glimpse of the task that harvesters in Spain and Portugal faced every season. The work was heavy labor and impossible to mechanize. As the hospital patients watched the team grapple with their saws and axes, Metcalf must have wondered, "Does this elaborate labor have a place in California life?" It was exhausting.

"Stopped in to get some cider at the Blaupers place and it helped," he wrote at the end of one long, hot day. As summer gave way to fall, his team spent more and more days cutting cork from large trees and gathering millions of acorns. The foresters selected for the adult tree's performance at a variety of elevations and on different soils, and studied the acorns from all angles, including moisture, weight, and nutrition.

In September Metcalf and his team hauled trucks loaded with slabs of cork to the San Francisco office of Crown Cork and unloaded them at the warehouse. He and his assistant were discussing the cork experiments when a special radio announcement overtook their technical review. "The big news of the day was the unconditional surrender of Italy at 9:30 a.m. Pacific Time," Metcalf wrote in his datebook. Italy's surrender allowed the Allies to fight Germany up the peninsula. German forces swept into Rome to hold off the Allies' advance, but the Axis empire in Europe had peaked. It was possible to imagine that Hitler might lose.

* * *

Charles Jr. was surprised by how well the industry had adapted to the strange bedfellows of the war economy. Behind Armstrong Cork's publicity campaign in the *Saturday Evening Post*, the company acknowledged that it was an unlikely defense partner. "The nation's leading producer of linoleum and cork insulations and bottle stoppers didn't seem to have much to offer to military buyers of guns and tanks and airplanes," an internal Armstrong account noted. In World War I, Armstrong had made itself into a producer of shells and camouflage (adapting its experience with floor coverings), and in the current world war it created a new Munitions Division that churned out 22 million shells, 4 million cartridge cases, more than 45 million square yards of camouflage, and nearly 11 million magnesium bombs. Military orders soared, from just over $500,000 a year in 1941 to $36 million in 1943 and $39 million in 1944.

Adapting to round-the-clock shifts, industry workers had to find new ways to get to the job. Crown and Martin arranged with Baltimore to have streetcar lines accommodate wartime workers. At Armstrong, when the bus from Strasburg didn't run late enough for her shift, one worker walked seven miles to the Lancaster plant for her 11:00 p.m. clock-in time. Armstrong's Pittsburgh plant workers raised carrier pigeons for the army. The company recycled carbon paper, and employees used fewer paper clips so the metal and paper could go to defense production. (Armstrong accountants counted ninety-five thousand clips used per month, a 40 percent decline since the war started.) While defense work brought opportunity, it also brought bureaucratic snarls that roiled Prentis and his team—like when the government froze prices on commodities, even though the costs of raw materials and labor rose. Armstrong management complained repeatedly to the Office of Price Administration, "which usually responded slowly and often not at all."

Prentis and Armstrong were trying to prepare for postwar production and consumer demand without compromising the ongoing war effort. Prentis gathered his managers and led a five-year planning exercise intended to anticipate postwar demand. They debated goals for 1948 based on where they expected the American economy to go. That included new materials for home construction and industrial uses, including new linoleums. They drew up plans for factories in Illinois and Mississippi that would be operational within two years of the war's end.

To plan for postwar staffing, Armstrong worked to make good on its commitment to loyal employees. More strategic than Crown Cork, Armstrong sent letters to all its employees serving in the military, inviting them to return to their jobs with the company after the war. By the end of 1945, more than twelve hundred former Armstrong employees had come back. Many others would opt to get more education and training with the GI Bill.

* * *

In January 1945, Woody Metcalf prepared to set out from Berkeley on a truck ride of more than thirty-five hundred miles across the country. He received a phone call from George Greenan on January 22: the truck had arrived and was being loaded with cork. They'd be ready to leave that afternoon.

Rolling with an eighteen-wheeler across more than a dozen states, the McManus Cork Project hit the road like a traveling circus, complete with the big-lettered sign on the truck, and arrangements for reporters' interviews and photo opportunities along the route. Those first months of 1945, as the war appeared close to ending, offered possibly the last chance to make a change in how Americans saw forests and how they planted trees.

Metcalf wrapped up affairs on campus; kissed his wife, Elizabeth, good-bye; and caught a ride across the Bay Bridge to San Francisco in the afternoon light. There a photographer was taking publicity photos of the truck being loaded. Metcalf and Greenan mounted the eighteen-wheeler and crossed back over the bridge in the dark. "Supper in Oakland," he wrote in his journal. They spent the night in an auto court in Modesto.

After a frosty night that left ice around the puddles in the parking lot, Metcalf and Greenan headed south and tuned the radio to KYOS for a program on cork planting and seedling treatment. In Fresno they loaded more acorns, and at Kearney they took on nine cartons of cork slabs. All the Bakersfield motels were full, so they continued into the Sierras, the "Bulldog" crawling up the steep grade at 10 mph. The next day they reached Crown Cork's LA plant. Reporters and a photographer showed up to document the road trip for the local press.

The cork crew loaded nearly eight hundred pounds of acorns in apple boxes and then the two men left LA for Beaumont, where they stopped at the San Gagorin Inn, and the air was cool with the scent of flowering almond trees.

In Arizona's arid landscape, Metcalf spied fascinating big *Carnegia* cacti alongside desert trees. They drove on to Wichenburg, where they bought a copy of that day's *LA Times*, containing an article about them. "U.C. Professor Helps Cork Culture in U.S." ran the headline.

> In a truck containing 12,000 cork oak acorns, Prof. Woodbridge Metcalf, extension forester of the University of California, George D. Greenan and Charles Gassmann passed through Los Angeles yesterday en route to government stations in Southeastern States, where the acorns will be left for future planting. Cork is almost impossible to obtain now from the Mediterranean area and there is great need for more extensive cultivation of the cork oak in America, Prof. Metcalf said. The United States uses about 300,000 tons of cork annually, 60 percent of the world's production, most of it heretofore having come from overseas. The acorns to be taken East were gathered by Prof. Metcalf and his associates on their trip from Butte County to Los Angeles County. With enterprising culture of the trees, Prof. Metcalf said, it is hoped that within 20 years sufficient cork may be produced to supply local needs. Today there are 4000 of the trees in California, half of them in Los Angeles County. Within the last three years 125,000 young trees have been planted in this State.

It was a dull article with little context. The LA reporter had scribbled down the basics. Against the backdrop of war, reporters grew lazy on everything else. Metcalf clipped the article anyway and stuck it in his datebook.

The next morning the truck reached Phoenix. It passed eucalyptus, radiata pine, tamarisk, oleander, and palms. Metcalf saw orange trees flourishing in the Salt River valley. They spent the last twenty minutes in a thick fog, on a highway slick with mud from recent rains.

In Phoenix, Metcalf and Greenan met with Crown Cork's top management: Charles McManus Sr. and Eva were there with Russell Gowans, Buell Jade, and Henry Costa. McManus had aged visibly since the last time Woody had seen him. And Metcalf missed Melchor Marsa, the guru who had been in the business forever and worked with producers in ancient Seville. Marsa had come to California and Arizona, rolled up his sleeves, and inspected moisture levels. He knew cork oak inside and out. Now he was in Portugal.

The old cork oaks they found growing in Superior were stripped, tested, and found to be good quality. The oldest grew on the Pinal Ranch between Superior and Miami, Arizona. The owner, D. I. Craig, traced the trees back to his grandfather, R. A. Irion, who received a gift of cork acorns in 1879 for subscribing to the San Francisco *Bulletin*. From a dozen acorns that his grandfather planted, one tree survived the extreme temperatures (ranging from 0° to 110°F) into the 1940s. It yielded nearly a hundred pounds of cork every harvest. The Pinal Ranch cork oaks had spread cork farther through the state. In Tucson, two cork trees on the campus of the University of Arizona dated back to Pinal Ranch acorns planted in 1921. (The largest of the cork trees on campus survived well into the twenty-first century and was named a Great Tree of Arizona in 2004. Steve Fazio, a plant science professor who worked at the university when it was planted, would take his children to show them the tree for years after the war ended.) Four more trees grew in Tempe, with several more in Litchfield and at the arboretum in Superior.

The event this time in Phoenix marked a new summit of home-front propaganda for the company. The cork-growing campaign had engaged thousands of Americans, distributed around 5 million cork acorns, and galvanized governments and civil society for environmental action. A photo captures the moment with McManus, the campaign's founder and funder, standing at a nursery in Arizona. Metcalf wrote: "Saw about 10,000 cork oak seedlings growing at Superior and brought here for distribution. Took pictures at Biltmore Gardens of Mr. McM and cork oak." Hundreds of Salt River valley residents received thousands of cork seedlings adapted for their desertlike conditions.

This was the high point of the McManus Cork Project. For six years the old man had bankrolled it all: the gathering of acorns, the nursing of seedlings, the climate zone research, plus so much state and national marketing. Giles Cooke had created a publicity machine that would make Henning Prentis proud: a system for generating local newspaper coverage and radio spots, engaging state and local civic groups and politicians, and championing a patriotic cause. The campaign came with appeals for actions you could do to win the war now and for your grandchildren in the future.

The economics of a pilot effort rarely look good. Changing an entire supply chain of a product is like cultivating a new energy source: a long shot that can take decades to break even. The McManus Cork Project yielded just over 32,500 pounds of cork in nine years. McManus couldn't say that this effort to implant cork in a new land had made a large dent in production, and it had cost hundreds of thousands of dollars of his own money.

But Crown Cork's wartime investment was paying off for shareholders. Cork harvests were piling up in Portugal, Morocco, Algeria, and Spain—and the US government was helping Crown to arrange for their transport to the United States as soon as shipping space permitted. Even Crown's tin can manufacturing was coming back to life after nearly expiring. The army's quartermaster was stepping up orders for what the troops required—massive volumes of beer. "Crowntainers" would soon be rolling out of factories in large numbers. The Highlandtown plant, which had shifted completely to war production, would soon return to civilian use. Like Armstrong, Crown Cork and Seal was starting to imagine life after the war. "We got through that all right," Charles Jr. said later. "It brought some developments. We cut the cork a little thinner."

Charles Jr. continued to represent his father at public events. Even as he placed seedlings in the hands of boys and girls across America, he wondered if those shoots would reach maturity. He wanted to believe the excitement of American-grown cork trees sown across the country, and a harvest of a thousandfold increase in American cork and families as committed to their trees as people were where the cork was a familiar native. But the mood in those days was not exultant.

In Phoenix his father grilled Greenan and Metcalf about California, the harvests in Napa, and the challenges of making harvests more efficient. They talked about the war and how the economics of their business would change—probably for the worse. Germany would fall within months, it seemed clear. The war in the Pacific could go on for much longer.

Nobody talked much about plastics, but a new age was dawning. The McManuses saw that cork was no match for the synthetics emerging from wartime research. Some of those synthetics had been developed by America's enemies: Armstrong was developing new products using materials research from Germany.

"There is no doubt that plastics will be used more and more in the construction of houses, automobiles and airplanes, and that more and more articles of plastics will be sold in bargain basements and in five-and-ten stores," Waldemar Kaempffert wrote in the *New York Times* in July 1944. Plastics confirmed that America was going through a chemical revolution. New techniques for freeze-drying produce were revolutionizing food culture, and synthetic paints and built-in lighting were changing living spaces. "When we begin to take for granted what looks like a glass tumbler but is actually a synthetic that can be dashed against a wall without breaking, something extraordinary has happened," Kaempffert concluded. Scientists were looking ahead to microwave ovens; clothes made of rayon, nylon, and other new fibers were already rolling out of factory mills. With synthetics, new cars would run more efficiently on tires that would be nearly puncture-proof.

During the previous six years while the McManus Cork Project grew, the secret Manhattan Project had also matured in the desert at Los Alamos, New Mexico. A few months after the Crown Cork promotional photo shoot in Phoenix that February, the Trinity Site near Socorro, four hundred miles to the east, hosted a preliminary test of a new explosive that would come to symbolize a whole generation. On a wooden platform, the staff loaded one hundred tons of TNT spiked with irradiated uranium, dissolved and poured into a tube. "The Gadget," as it was called, exploded with the force of around twenty kilotons of TNT. A handful of the country's best physicists were on hand as witnesses. In the predawn darkness, the explosion lit the ring of mountains around the site—first purple, then green and white, "brighter than daytime"—for nearly two seconds. The El Paso newspaper noted a "blast was seen and felt throughout an area extending from El Paso to Silver City, Gallup, Socorro, and Albuquerque."

In the desert, the future was becoming clearer, like Phoenix the city, emerging from fog in the truck's windshield. The Phoenix newspapers covered the cork project, noting that the project had distributed cork oaks in Arizona since four years before, when McManus found a cork tree "growing on the Arizona Biltmore grounds, unrecognized by anyone as a cork oak. . . . He hopes the acorn distribution will result in the starting of millions of trees."

Over lunch with the Crown Cork managers, Metcalf talked through the schedule of the long journey to Baltimore that still lay before them and beyond that, about how the campaign would look in peacetime. McManus Sr. joined them and the discussion continued over dinner, then meandered into more personal directions over glasses of Benedictine and brandy before Metcalf retired to the Hotel Arizona for the night.

Metcalf and Greenan resumed the trip the next morning, driving east through fog up to Superior, to the Coolidge Dam and across to Safford. The mountain frost gave way to a new landscape that enchanted Metcalf with its creosote birches, mesquite, and short-stemmed yuccas. He connected more with the flora than with the residents.

The rolling corkathon crossed the Rio Grande and reached El Paso. Metcalf and Greenan checked into the Del Camino auto court on Highway 80, where they pored over a map and sketched out their 640-mile route across Texas. Near Sheffield they passed an oil field with its blazing plume. They passed junipers, scrub live oaks, pecans, and more juniper woods, where sheep and angora goats grazed. They crossed the Plano River, on to College Station. At Texas A&M, foresters gathered to see Metcalf's slides of cork oak and learn about the tree that could shore up America's defenses.

The two men drove east to Beaumont, crossed the Mississippi in a driving rain, and reached Lake Charles, Louisiana, late the next morning, delivering one thousand pounds of acorns to the Louisiana State Nursery. The landscape turned to forests of pine, cypress, and oak with Spanish moss. "Miles of sugar cane and rice fields."

"Bad Cajun coffee," wrote Metcalf, fueled their ride through sugarcane country to the mighty Huey P. Long Bridge into New Orleans. They left the truck to be serviced at a Gulf station at Canal and Claiborne and walked to the port for a look around. Metcalf bought pralines to send home, and then he and Greenan took a bus out Esplanade to City Park. There a forester from Louisiana's agricultural experiment station gave them coffee and complained about how the cork oak program was being managed in the South. State politics were getting in the way. No matter how wonderful the scene of crowds at the state capitol on Arbor Day, the reality in Louisiana was that few seedlings would ever reach harvest age.

That night, amid the temptations of Bourbon Street, Metcalf wrote postcards home. "Eternal vigilance is the price of liberty," warned Thomas Jefferson from the margin of his datebook.

In Mississippi the cork Bulldog passed lumber mills in the rain and dropped off two boxes and three sacks of acorns for another state forester to distribute.

As they sat down to dinner in Meridian, they heard news from the Pacific: "Radio reported that MacArthur troops retook Manila," Metcalf wrote. MacArthur had made good on his oath three years before, returning to the beaches of Luzon.

Domestic affairs were less encouraging. Driving through the South, Metcalf was startled by the poverty of the black communities they passed. "Colored shacks very poor," he wrote.

In mid-February the truck made its way from the Great Dismal Swamp at the North Carolina line to Washington, finally reaching Baltimore, where the team was welcomed by Giles Cooke. Metcalf and Greenan changed into fresh clothes in the Lord Baltimore Hotel and enjoyed a "fine supper" with Cooke at Milton's.

"When a company truck pulled into Baltimore today it marked the end of one of the most extensive known such seed scattering ventures," wrote the *Washington Post*, noting that more than a million acorns had been distributed for planting across 10 southern and western states. "Charles E. McManus . . . has far exceeded anything Johnny Appleseed could have dreamed." The article added that it would take 10 million producing cork trees to meet America's supply needs.

While the last of the California cork was unloaded at the Crown Cork and Seal plant for analysis, Metcalf explored Baltimore by streetcar, with a visit to Fort McHenry.

At the Crown Cork factory, the research team, including Cooke's colleague Victor Ryan, met with the McManus brothers to discuss the cork program at length. Metcalf took in the seventeen-acre factory complex and admired its cafeteria. That night after dinner, at Cooke's home in Baltimore, Metcalf projected slides of California cork oaks onto the wall of the dining room. Pictures of large oaks flashed like magic lantern images of a possible future.

Charles Jr. tried to sort through, as his father stepped back, how the company should move forward. The last months had gotten even busier for him, with new duties. He was one of a group of businessmen named to chair committees for Maryland's postwar economy. A national organization arranged by President Roosevelt's secretary of commerce began holding meetings about shifting industry to peacetime operations, and Junior found himself part of that effort at the state level. He headed up the Maryland group from glass, paper, containers, and metal companies—the usual suspects. The VP of the Martin Company headed up the aircraft group, A. G. Decker from Black & Decker led the machine building and machine shops group, and the manager of Bethlehem Steel handled the committee for heavy steel and fabricating industries. The committees had two goals: high employment and a speedy return to peacetime status quo. Simple goals; tedious meetings.

Newspapers tracked growth in retail sales, predicted shortages ("Meats may virtually disappear from civilian markets by midsummer"), and detailed the story of four army privates who had faked their transfer papers. The GIs had signed themselves out of the war in Europe, hopped a troopship back to Baltimore, and proceeded to Washington, where they visited the Pentagon to write their own discharge notices. The Sunday *New York Times* of February 11, 1945, contained a short article describing how a black American fighter squadron based in Italy had returned a $1,000 donation from a Detroit union local, requesting that it be used instead to improve race relations at home by establishing an annual prize for "the person or group contributing the most toward racial good-will." The squadron's commander added that his men had made their decision from a desire for "understanding and advancement of the Negro."

The next day McManus said good-bye to Metcalf, until the campaign moved into its next phase. The forester boarded a train at Baltimore's Penn Station and headed back to California.

* * *

Charles Jr. booked a passage for Europe right after the jubilant relief of the victory celebrations subsided. His assignment from his father was to assess damage to cork sourcing and shipping channels, set Crown Cork's business in order for resuming shipments, and get operations in Europe back on a commercial footing. With Melchor Marsa in poor health, McManus Jr. needed to check on the Spain and Portugal operations. He felt conflicted about the trip and anxious about leaving Mary, who was about to have their third child.

Like his mother, Charles Jr. documented his experiences in a journal every day. And he wrote Mary almost daily, numbering the letters so she would know if any got lost in the ocean crossing.

Before setting off, he trained up to New Jersey to seek guidance from

Herman Ginsburg. Exchange Place bustled with businesses making up for lost time. McManus braced himself to find a Europe drastically different from the one he had visited before the war, even from his and Mary's time in England a few years before. Ginsburg had worked with Melchor all through the war, so he knew where to look for skeletons, so to speak. The Algerian plant, for starters. There were news items about Algerian labor disruptions and independence protests. Had these closed the plant down? There were many rumors.

Herman invited Marsa to join the discussion, so Charles could learn what he could from the master. Melchor was living back in Brooklyn, not looking well. Still, Charles's father wouldn't accept Marsa's resignation. The old Spaniard had every right to rest, yet he was making preparations to take his family back to Lisbon now that peace was restored. Melchor came and the three of them had lunch.

Charles considered that Herman and Melchor had worked together for over ten years, through the most difficult times in the century. As they spoke at lunch, Charles scribbled lists of whom to contact, what to ask, where to look. Herman expressed confidence that the cork supply for next season was assured. But what Junior would find in London, or a just-liberated France, or Franco's Spain, nobody could say.

Marsa's observations on Lisbon were astute as always. Junior left the meeting feeling the support of these wise veterans but also concerned that he had little ammunition for navigating an uncertain, war-ravaged terrain.

On the ship, Junior read and heard from other passengers about their expectations and fears going back to Europe. It was an emotional passage. Through the war, the *Queen Elizabeth* had served as a troopship, and it had just been refitted for a new wave of commercial passengers.

The rubble and damage in London shocked him, but it also prepared him for what he'd see on the continent. Antoine Leenaards was rebuilding the company operations in Belgium. When Junior reached Paris, he cabled the plant manager in Algeria to let him know of his plan to visit Algiers. Mr. Joann replied immediately: he would take the train up from Marseille and meet McManus in Paris instead. Junior could just imagine what that meant: things in Algiers were a mess.

Joann had been the Algiers manager for years, dealing with Vichy officials to get his shipments out. The plant made only wine corks, and through the war it had sold to both sides. Now that the French had retaken control of their colony, they were prosecuting Joann for dealing with the enemy. Even coming up to Paris proved difficult.

Joann told McManus that he wanted to buy the factory. Junior asked what terms he wanted, what they could expect from it. The man had a haunted look. When they shook hands on their way out, he already seemed absent.

The McManuses agreed to sell him the Algiers factory. Joann had run it like it was his anyway, and it added nothing to Crown Cork's bottom line. The little cork it produced almost wasn't worth loading onto the ships.

Algerians had hoped, after being liberated from the Germans, that they would gain their autonomy. But little changed under the return to French rule. In fact, the French expanded land ownership by French nationals and displaced Algerians even more. Finally, the simmering resentment erupted in violence, in small towns and later in Algiers. The unrest continued to grow until revolution and independence came in 1954.

Charles gained just a glimpse of a rebuilding Europe in his short visit, something like peering through the spectacles of an eccentric manager. He jotted notes in his diary and letters to Mary—about the hotel rooms, the staff who had survived the war, those who hadn't, his desire to be home. In the years right after the war, he would return with Herman and Bobbie, traveling like family—teasing each other, trading stories, going to the theater together. Between meetings with associates in Lisbon, Charles the corporate scion would take walks by the rock coast. Once he caught a ride near Cascais with a local couple to escape a downpour. They took him to a restaurant built inside a cave and shared a lobster meal. It turned out the couple owned a cork grove in Évora, sixty miles inland.

Charles was anxious to be home when the baby arrived. They named her Eva, after his mother.

Many years later, young Eva met some of her father's business colleagues, including Ernesto Mas, while traveling in Spain. During the war, Mas had worked for Melchor Marsa and Crown Cork in Lisbon. Mas told Eva that her father and grandfather had offered to bring him to the States to escape the war. He didn't take them up on the offer, but he always remembered their generosity.

The McManus sons were handling more and more of the company

management and continued to shoulder a large role in the cork project. Giles Cooke guided the machinery for plantings and publicity—methodical work, possibly a little bloodless. The greenhouses churned out seedlings, but not with less urgency than before. McManus had more trouble getting responses from politicians' offices. The congressional election of 1946 was shaping up to be a bruising one. Truman didn't have the support that FDR had enjoyed, especially when he bungled the response to labor strikes by autoworkers in 1945 and mine workers in early '46. Republicans were turning him into a joke. "What would Truman do if he were alive?" they said. Or, "To err is Truman."

Charles Jr. oversaw arrangements for Maryland's Arbor Day festivities that spring in a different atmosphere than before. Governor Herbert O'Conor was preoccupied with his bid for the Senate, one of many Democrats nervous about the fall election. He was eager to join the popular event and its publicity, and leery of hard commitments.

Charles and Walter attended the Annapolis event as special guests, and Governor O'Conor presided. On a platform built for the occasion, Junior spoke into radio microphones about the origins of the McManus Cork Project—the decision to reduce US dependence on foreign cork sources, the tree's resilient bark, the tests of the California cork, and findings of "excellent quality."

"Realizing the tremendous value of having this essential tree grown in our country we established a Cork Project," he said. "The purpose of this project is to provide in the United States a domestic source for at least a portion of the nation's cork requirements and at the same time add to the country's natural resources." He praised foresters across twenty-two states for the large plantings done and the pilot efforts in other states.

A band from Baltimore City College launched into inspiring tunes, with more than four hundred schoolchildren gathered. Scores of garden clubs and 4-H members listened to how "pioneering of the planting of the Cork Tree in the U.S.A. is constructive and an assurance of the progress of our civilization."

The message had shifted. Since the war's end, the Arbor Day speakers talked less about American self-reliance and more about jobs, and the cork-growing industry as a welcome addition to Maryland's economy. "The forest is unquestionably our most important renewable natural resource," the governor said. "When the last shovelful of coal or ore is taken from a mine, that mine ceases to exist as a natural resource. . . . The forest, on the other hand, by natural growth, is able to renew itself indefinitely if given a chance."

The governor disparaged clear-cutting and other destructive practices as "the worst sort of folly." Then he posed with the McManus brothers beside a seedling that represented the future. The April day was brisk and the grass was just greening.

Junior was concerned. The war's end was a huge relief, of course. But with the common cause gone, many divisions among people now surfaced. He found himself blindsided by questions about Crown Cork's wartime contracts, profit margins, and lack of competition—even the motives behind the cork campaign. Those questions didn't come up often, but when they did, they felt like a punch.

Armstrong's Henning Prentis was on the warpath for Republican candidates in the midterm election. Prentis was what Charles Jr. called "a real tight operator." Under his management, Armstrong didn't let wives join their husbands at bottling conventions, unlike everyone else in the industry. Prentis railed against any economic planning exercise as "socialist" in speeches across the land. In a planned economy, he warned, the individual would become a "servant of the state," the opposite of the Founding Fathers' intention. That spring, at Middlebury College in Vermont, Prentis's commencement speech cautioned that "when the Government becomes the primary source of capital funds, national socialism automatically displaces representative democracy." The government's role in the economy had indeed changed. It had accounted for one-tenth of gross national product in 1939; by the war's end, it accounted for half.

Less predictable were former business partners and politicians. Cork got caught in the crosswinds. Even the seedling they planted that day on the Annapolis grounds, steps from the state capitol, suffered. "I'm sure that was sabotaged, because all of a sudden it just died," Charles Jr. said later. "Someone poured gasoline on it or something. It had to be that, the way it died."

Ultimately the public, too, stopped caring. They didn't sustain the effort. "It takes too long," he'd say. "People aren't interested."

* * *

McManus Sr. saw the returning GIs coming through Baltimore's Penn Station. He felt he no longer understood many things about America.

He accepted the changes, but they came much faster than they had after the Great War. Now Crown Cork lost government contract renewals to smaller companies that leveraged new materials and manufacturing processes. The ambition of the government's recovery plans were staggering. For Europe! General George Marshall's ideas about rebuilding Europe were breathtaking—far beyond the most ambitious defense policy. In the *Times* the previous fall, Marshall had published a sobering revisionist view of how the war had been won: he said that the British and Soviets, by showing resolve during the war's darkest days, had shielded Americans from a war on US soil. Marshall credited the Allies' victory not so much to American resolve than to the enemy's missed opportunities. The North African campaign, along with the Russian victory on the Volga, had marked the war's turning point, he wrote.

McManus recalled that campaign, when Allied forces had crawled through cork forests to liberate Algeria. He remembered late '42, when everyone was on tenterhooks about Algeria and the Iberian Peninsula hung in the balance, isolated by Axis-controlled Europe and German forces in North Africa. And he thought of Melchor Marsa.

Besides snooping around Lisbon for the OSS, Marsa had been invaluable to the Cork Project. He knew best the growing patterns and the harvest methods. If McManus joked about Marsa's advantage growing up in Catalonia with roots instead of feet, the truth was that it was hard to imagine the epic effort without him. Now McManus could picture, on his train rides through Maryland and across to Arizona, someday passing through a savannah landscape of cork groves like the ones that enchanted him in Europe.

The September 1940 blaze at the Highlandtown factory remained an unsolved mystery, like the Hercules Powder factory explosion that same month. The FBI had closed the file with no criminal prosecution. "One convenient thing about an explosion," an op-ed columnist wrote in the *New York Times* about the Hercules blast, "is that it ordinarily destroys the circumstantial traces of how it happened." The writer noted sabotage cases from World War I that were proven only decades later. In the case of Crown Cork, officials questioned whether fire extinguishers had been tampered with, perhaps by a disgruntled employee. McManus considered it simply another in the string of factory fires that marked the industry's history. Charles Jr. had his own theory: workers on break had been smoking (against rules), and someone's cigarette butt set off the eighteen-alarm disaster. Against that fall's drumbeat of news of Nazi intrigue, xenophobia, and war in Europe, what probably started as an industrial accident ignited public fear and government scrutiny of their entire industry.

As McManus handed the redcap a twenty and boarded the train for Spring Lake to join Eva in New Jersey, Baltimore seemed to be slipping away from him. The fingers of high-rises stretched up from the low-slung skyline of his hometown. He could hear Eva ask mockingly, *Is that a city?*

He was leaving the boys in charge of Crown Cork, but something nagged at him.

Aging was hard for Charles and Eva. She was often in a bad mood these days. Their trip to Arizona would be a good antidote. The train ride out took five days. Then the desert air and sunshine lifted the spirit like a new world. The cork trees in Arizona were growing well, according to Greenan and the forester Metcalf. If half the plantings they had mapped across the country took root, that could justify the investment. Crown Cork's board and shareholders might not agree, but McManus had used his own money. The board would follow his lead.

That was what nagged at him. Charles and Walter, despite their commitment, didn't yet have credibility with the board. They needed more time to get settled in, then he would step back.

The plantings in Arizona anchored him in a future he could fathom. Everyone else seemed transfixed by this other one, the sudden nuclear age. Science held brilliant and terrifying prospects. Cooke's latest article, in the *Journal of Chemical Education*, already looked out of date compared to the issue's cover article on the emergence of alloys. *Science Illustrated* had a shiny futuristic dome on its cover, where Cooke's piece "Tree with a Future" jockeyed for attention with articles on the atomic age, warmth cubes, radar, "Surgery with a Trowel," and the "fastest thing with wings."

Cooke told McManus that he himself was dabbling in research on synthetic resins.

It would take decades to replace cork as a sealant—McManus felt sure of that. And in beverage bottling, its quality remained unmatched. You could taste the difference between a cork-sealed bottle and one sealed with plastic. There was no question in his mind.

* * *

In late May 1946, when he and Eva returned from their annual trip to Arizona, McManus checked into the New York Hospital. He had been feeling poorly, and the desert climate had not done its usual magic. In the hospital for several days, he considered where things stood for his family, for the company, and for himself.

He had a visit from Herman Ginsburg. McManus couldn't say that he remembered Ginsburg's first job in the mailroom, but he was proud of how the young man had advanced in the two decades they had worked together. They talked about world events. On one hand, there was a new United Nations. On the other hand, a new kind of war without battles was already brewing with the Soviet Union. A cold war. *Always something new, eh, Herman? We'll see what happens.*

Ginsburg left the hospital room and walked down the hall, hoping next time to raise the subject of plans for the new plant in South America.

But at the end of a week, McManus was slipping. In a blink, his parents, Eva appeared—then the two babes they lost as infants, then the boys. On a ship, or rather walking from the car to the forest's edge, where the stripping crews were at work, the sunlight pink on the trunks. Then he was entering the Highlandtown plant, and Giles Cooke was saying something, pointing to the slabs on the truck. "It's just—" Cooke was saying. McManus couldn't make out the words.

Eva was leaning over him. "You," he said with relief. And breath left him.

The mind that spawned an empire of bottle caps, and that envisioned a nation under oak canopies, expired. But those empires and visions were already shifting in a new modern age.

Chapter Nine

# COLD NEW WORLD

## · Marsa/Ginsburg: 1945–1961 ·

Very often in history . . . the fate of nations has depended on the success of a small anonymous group, who have taken their lives in their hands to carry out an espionage mission to save their homes or to serve an ideal.

DONALD DOWNES, *The Scarlet Thread*

Families are always rising and falling in America.

NATHANIEL HAWTHORNE

When Gloria Marsa saw colorful ads on the subway, with a soda bottle or a floral pattern, she was reminded of Lisbon's bright hues. In the months after the war ended, she allowed herself to get excited. For years it seemed impossible that Europe would survive Hitler and the Nazis, but now that cloud was gone. She was twenty, ready to graduate from Packer Collegiate, and was preparing for a new start in Lisbon with her family.

With graduation approaching, she and her parents packed up their things to ship to Portugal. She was eager for European life again and a new phase, a career in the new peace. They would return aboard a ship like the one that had caused such anxiety five years before, when she had looked up at the Lisbon skyline uncertain if she would ever return.

In the early months of 1946, though, her father's breathing grew more labored and his health remained fragile. The doctor said the problem was emphysema and recommended that Melchor not travel yet. They would stay in Brooklyn for a time until he improved.

The months stretched on, and her father didn't seem to be getting better. Gloria saw that their plans were quietly shifting. So she relaxed her expectations for a broadcasting life in Lisbon. She applied for jobs in New York and found a position with an export company on Wall Street that sold leather goods.

Gloria watched for an improvement in her father's condition, but there was none. He didn't go out much, and when he did, he quickly grew tired and short of breath.

When the family learned that Mr. McManus had died, the news startled her. McManus was only two years older than her father. It came as a shock to them all. Gloria accompanied her parents on the ride across the river to Madison Avenue near Central Park, and they paid their respects at the wake held at Campbell's Funeral Home. There were many colleagues and friends from Crown Cork and Seal, including Herman and Bobbie Ginsburg and Mrs. McManus's brother, Leonard Olt. And surrounded by a crush of mourners were the McManuses—Eva and sons Charles and Walter and their wives. Gloria remembered the sons from their visits to Seville and Lisbon. Gloria's parents offered their condolences in a formal way. Charles Jr. seemed touched.

So was her father. The Marsas and McManuses had spent two decades in a shared enterprise. They had survived the Great Depression and a second world war and had played their parts in the victory. Melchor Marsa had served quietly, while some men who had been in intelligence received honors, like the Park Avenue ad executive who received the Legion of Merit.

Gloria didn't know that part of her father's life. In the taxi returning to Brooklyn with her parents, she saw him exhausted after visiting with so many people whose lives had become prosperous. Her father had given so much to others, to the company. He had been important to its survival and had shared his wisdom, both in Portugal and in the United States with the tree-growing program. It was troubling to see him now in a taxi, struggling for each breath.

Years later, a Crown colleague from Lisbon came to visit Melchor Marsa in Brooklyn and pay his respects. "To this day," Gloria heard him tell her father, "we never fail to have a meeting of all the cork people where they say, Now what would Mr. Marsa have done?"

* * *

After the McManus memorial service Herman Ginsburg took the train back to Jersey City. The Crown office at Exchange Place had grown to include a corporate assistant, two accountants, a couple of clerks, and a secretary—still a small staff for managing far-flung international subsidiaries. Herman and Bobbie traveled twice a year, once in the spring on a first-class ocean liner to Europe, and a second trip in the fall to Latin America.

With the passing of McManus, Ginsburg gained more standing in the company, and more autonomy from Baltimore. He prized that freedom, a colleague said: "He worked very hard in his arcane way to make sure that he maintained that independence."

Ginsburg rarely talked about his career goals or his feelings even with his family. He often took his lunch with associates at an Italian restaurant, Bruno's. He brought his own ideas to management and acted on them. When he hired managers for Crown factories overseas, Ginsburg placed an emphasis on their knowledge of local conditions and relationships. He hired local talent at a time when the standard practice was to hire American expatriates. In that, he was far ahead of standard practice; he had learned from people like Melchor Marsa.

At one point Ginsburg's niece, Connie, asked him, "How was it possible for you to go from being the oldest of eight children in a very poor family to a high official in a corporation?" He replied that when he was a boy, the family would gather on Sundays. Ginsburg's father would ask the cousins who came, "How are you doing?" And they would always answer, "Making a living." Herman found that depressing. "I knew I didn't want to do that," he told Connie, "just go that far and no further, having that kind of life." He wanted to go further.

After the war Ginsburg commuted daily from pastoral Mount Freedom, New Jersey, to noisy Exchange Place. At home he kept a garden, where he tended roses and took pride in their exuberant blooms. That was the picture that few people saw: Herman in the garden on a summer morning, happy with his plants.

His friends were mostly Bobbie's friends—musicians and artists. He and Bobbie continued to see the Marsas. Often Herman visited Melchor at home in Prospect Heights. But Bobbie was their social director.

The couple took special interest in Bobbie's niece and nephew. With no children of their own, they supported the children's education and paid for Connie's music lessons. "He played a big role in my life, what I was able to do," she said later. "He was very modest. He saw himself, he said, like a very good grocer. 'I keep track of what I have, and I know what I don't have. There's nothing particularly unusual about what I do,' he'd say."

Ginsburg saw Crown's business rebound after the war. By 1950 global cork production was higher than it had ever been. And even with the wealth of synthetic materials emerging, cork was still a factor in national security. In 1951 the Central Intelligence Agency (which replaced the OSS) produced a classified report that brought cork's value into the Cold War. Intelligence sources detailed shifts in cork production in the new geopolitical landscape and revealed that the Soviet Union had launched research on growing cork in its territories along lines similar to the McManus Cork Project.

"The increased imports into the Soviet Bloc during 1950 reflect a growing need for cork," the report said, "and possibly for stockpiling." Despite synthetic advances, it noted, "There are no completely satisfactory substitutes for the principal industrial cork products, particularly gaskets, oil immersion friction drives, friction clutches, carburetor floats, washers, grease retainers, textile cots, grinding and polishing wheels, and certain types of insulation."

The CIA found that demand for cork was highest in the United States, the United Kingdom, West Germany, and the Soviet Union—with the United States still taking about 40 percent of world production and the USSR accounting for about 25 percent and growing fast. Soviet imports of cork had increased by more than 50 percent over the 1945–1949 average. Cork had also spread to industries in Japan, China, and Korea.

Soviet demand for top grades of cork, with apparent disregard of price, was pushing up the global price. The behavior supported a CIA theory that the Soviets were scaling up their cork imports for defense uses. The Soviets experimented with a special type of cardboard to replace cork gaskets and seals, but it failed to hold up. Cold War powers still needed cork. The CIA report placed cork's value in war on par with "critical materials" like industrial diamonds. It stated, "As a result of the strong world demand for cork, caused to some extent by fear of another world war, production and exports in the calendar year 1950 reached record levels of about 353,000 metric tons produced and 348,000 tons exported."

The Cold War report reached back into history. Russian attempts to establish cork plantations in the Crimea and Caucasus went back to 1819, followed by efforts in Georgia in the 1890s and Soviet plantings in Azerbaijan by 1930. Stalin in the 1930s had made plans to systematically enlarge the area of Soviet cork plantations by more than 6,000 acres a year, with the aim that by the late 1950s the Soviet Union would have 185,000 acres of mature cork forest, an amount that Soviet planners calculated would supply all of their domestic cork needs.

Stalin, however, was no match for cork's demanding nature. Many hundreds of acres of plantings withered from winds, poor soils, pests, and diseases. Despite an additional 2 million-ruble investment (approximately $59 million) in research and plantings, the Soviets could not beat the limitations of cork's natural range. By 1949 their plantations had shrunk to just over two thousand acres.

So Stalin needed to import cork from Portugal and Spain. Franco bartered Spanish cork for his country's needs. As the Cold War deepened, trade tensions escalated in the 1950s. At one Moscow dinner party at the Indonesian embassy, the Soviet defense minister made light of the competition with the West. As he raised his glass to the Soviet willingness to fight if it became necessary, a cork exploded from a champagne bottle. "Let's use those instead of bullets!" he toasted.

Several months later, France and the Soviet Union reached a trade agreement increasing the flow of goods between the two countries by more than half, including increases in cork.

Giles Cooke pressed the fight for American cork into the Cold War. In Norfolk, Virginia, at the headquarters of the Fifth Naval District, Cooke spread the passion for domestically grown cork among naval supply officers. At 8:00 p.m. sharp on November 14, they gathered at the Sears, Roebuck Community House to hear Cooke speak. "The national importance of Crown Cork and Seal is known to all supply officers," noted the memo from their commander about this excellent opportunity to learn about the operations of cork.

The main trade competition brewing was not for cork but for petroleum. The intensification over control of another natural resource followed a familiar pattern: America saw an increased dependence on a key resource for its economy and worked to secure continued access. Before the war, the United States had had 10 percent of the Middle Eastern oil market; by 1950 that number had jumped fivefold: the United States was importing half of all Middle Eastern oil. The petroleum economy was growing globally. Oil and gas had accounted for one-third of the world's energy in 1929; by 1952 that figure was up to two-thirds, climbing to more than 70 percent in the 1970s.

* * *

Gloria Marsa continued working with the leather export business on Wall Street, becoming invaluable to its bookkeeping and sales operations. Her father's health problems did not abate. He rarely got out of the house and no longer sang along with the opera on the radio. But he stayed deeply engaged with the world and his family. He kept up on developments in composition cork and Portugal's progress in cork technology, peppering visitors with questions. Gloria married and moved to Mexico with her husband, coming home often to see her parents.

Her father's last months were filled with doctor's visits; he spent the time tethered to an oxygen tank. The family changed the large cylinders four times a day. Gloria told people that his thirst for oxygen was a sign of his zest for life. "He's spent the oxygen that most people use in two generations," she'd say, "because he put so much into life." His hollowed cheeks and the shadows around his eyes were painful to see.

In March 1953, after an extended bout with emphysema, Melchor Marsa passed away.

The Fairchild Chapel on Atlantic Avenue was filled with a river of people coming to pay their respects. After the first day, the funeral home manager moved Melchor's casket to a larger room—the flower arrangements were overflowing. The next day they had to move him to a still larger room. Flowers surrounded the casket on all sides, and mourners filed through and consoled the family.

Gloria was with her mother receiving condolences when someone asked her to step out for a moment. Puzzled, she went out and found a man she recognized, her father's old business associate, Joseph Esposito, who ran a paint and varnish company. He was visibly upset. She asked him to come in to pay his respects to her father and mother, but he begged off. "Your father was such a dynamic and tremendous person," Esposito explained, "I cannot see him lying in a coffin. I just couldn't." He wanted to preserve the vigorous image of Melchor Marsa in life.

* * *

American business was becoming more globalized, and the international division of Crown Cork was the only part of the company turning a profit. The Jersey City office had more work than it could handle.

By the early 1950s, the Cold War had turned colder. Within the United States, McCarthyism created hysteria about the communist threat to American life. Every day Joseph McCarthy's Senate hearings featured another grilling. People gave up names of former friends. In this toxic atmosphere, when it could do the most damage, Herman Ginsburg's former life with his first wife resurfaced.

In pursuing every lead to root out communist activity, the FBI contacted a former communist organizer, who gave them Ginsburg's name. After testimony before McCarthy's committee and interviews with the FBI, Earl Clifton Reno, a former union organizer in Baltimore, identified Herman and his two brothers as Communist Party members in Baltimore in the early 1930s. Pressed about communist activity of that time, Reno told the FBI that communist leader Earl Browder had told him that the Ginsburgs had contributed around $3,000 a year to the party. The FBI pursued interviews with Lillian in January 1955. Agents called on Herman that spring, demanding his cooperation.

The FBI file on Ginsburg shows that in that period of the 1950s, when Crown Cork's plant in Baltimore was still considered a key facility for national security, Ginsburg became the focus of a disturbing investigation and series of interviews with agents. They interviewed his coworkers, the administrators at his alma mater at the University of Maryland Law School, his Baltimore associates, and his local postmaster in New Jersey. Ultimately the FBI left him alone. In the wake of that episode, a rattled Ginsburg became even more careful.

A few years after that, Ginsburg met W. Michael Blumenthal, a young Princeton professor with an unusual story. He and Ginsburg got to talking when they were both on the *Queen Elizabeth*, returning from Southampton to the United States. They had run into each other on a train going to the ship and had recognized one another from a previous encounter. As Blumenthal later explained: "My interests at the time—studying how US multinationals organized abroad to maximize their competitive advantage—had involved interviewing a number of US chief executives in their home offices. Among them had been a most unusual man with decidedly unorthodox views, completely self-educated but highly intelligent and successful, who ran a company manufacturing, of all things, 'crown corks'—or bottle tops—in a dozen countries around the world."

Blumenthal's study of CEOs examined decision making within international corporations. Specifically he wanted to know: Did a corporation's overseas managers hold the same views about decision making as the CEO? Blumenthal interviewed the heads of Esso (later Exxon), Ford, and other leading businesses and asked to speak with their managers abroad. Most agreed to let him interview the overseas staff. Ginsburg, unlike the other CEOs, declined. They hadn't seen each other since that one interview, until they shared a cabin in the train to where the *Queen Elizabeth* was at dock. Blumenthal continued: "On the first day at sea a surprise invitation had come down from [Ginsburg's] first-class quarters to join him for tea, followed by a challenge to what turned into a nonstop string of chess games stretching over the next several days."

Over the chessboard in his first-class cabin, Ginsburg heard more of Blumenthal's story. Born in Germany, Blumenthal was living with his family in Berlin after the 1929 crash, and the family business failed due to poor management. That business failure, which left them without means to relocate, very nearly killed the family when Hitler came to power. From that time on, young Blumenthal felt the need to understand the gears of business decision making.

As a teenager, Blumenthal had to get his family away from Nazi Germany. By 1939 their only option, along with a handful of other Jewish refugees, lay halfway around the world in war-torn Shanghai, the only place that offered them safe haven. After the war, Blumenthal made his way from China to the United States, and through determination and brilliance, he got into Princeton for graduate school in 1951. With his PhD, he earned a tenure-track position in the university's Economics Department.

Decades later, his career would lead him to the highest corridors of power—the president's cabinet and the position of US treasury secretary.

Ginsburg gleefully spent two days beating the young professor on the chessboard amid their conversation. Afterward Blumenthal recalled, "I thought I was interviewing him but he was really interviewing me."

Ginsburg asked him, "Tell me, why would an ambitious young man like you prefer to sit on the university campus and study what *others* do rather than testing his mettle by doing it himself?" Bottle caps might sound boring, he admitted, but the challenge of building a profitable global enterprise was the key. Did the young man know about corporate pay?

Blumenthal took the bait. Back in Princeton, he told his wife about the odd exchange. "You know, I think this guy offered me a job," he said. "What would you think if I went to work for him?" They agreed that it was an exciting opportunity.

Blumenthal followed up later with a visit to Crown Cork's office at Exchange Place. Blumenthal got right to the point: "Herman, about that job you offered me—" Ginsburg appeared to be surprised.

"Job? I didn't offer you a job," he said. "You mean you're *applying* for a job!"

Ultimately, he offered the young professor a job at roughly four times the salary he was getting at Princeton.

Crown Cork International proved an opportunity for Blumenthal to delve into all aspects of international operations and negotiations, with travels to the source of the materials in Portugal, Spain, and North Africa and to processing operations in Brazil and beyond. "It was a major jump from the Princeton faculty to this job. But it was tough. . . . I had a lot of world experience from the Holocaust and refugee years in China but no business experience."

Yet Blumenthal quickly saw that despite the scope of his work, he was on a short leash. Ginsburg had created his own quirky domain.

"He treated me like a son to be trained," Blumenthal observed.

One of his first assignments involved the delicate art of personnel management. With minimal background, Blumenthal was sent to England and the company's largest subsidiary. His task: fire the managing director.

"For a thirty-one-year-old kid to go over there and negotiate the golden handshake," Blumenthal said later, "was just madness." In British law, the agreement was called "compensation for loss of office." In this foreign terrain, the young man was up against the managing director, a gentleman who rode around in a Bentley with a chauffeur. But Blumenthal navigated the bureaucracy and the delicate egos, reporting back to Ginsburg on a daily basis by cable.

Ginsburg grilled him for the tiniest details about his conversations, analyzing each man's motives and attitudes. The Crown strategy was to reduce risk generally, with flexibility for on-the-fly decisions. When

Ginsburg was persuaded that a risk was low, he could be impulsive. In some respects, Ginsburg resembled McManus Sr., who at the Santa Anita Racetrack had made an impulsive bet on a 58-to-1 long shot.

Blumenthal became Ginsburg's eyes and ears, traveling to subsidiaries and helping to manage them. "There were times when I could have choked him to death," Blumenthal admitted. "Everything I told him he would double-check and triple-check." Ginsburg demanded details that seemed unimportant: Where did this conversation take place? Who broached the subject first? Do you think he was honest? Blumenthal recalled: "He liked those details because . . . he tried out of that to get a picture of what people were thinking. With me, he opened up a little bit, but he was considered in a way to be warm but unapproachable. He always inquired about the families of the people who worked for him. He always took presents for the children. He always wanted the wives included. . . . So Herman was considered to be a tough guy but a good boss whom you had to keep happy."

On one trip to Portugal, Blumenthal rode along into the countryside east of Lisbon. "I remember going out into the cork forests and looking at these trees where the cork came from," he said later. He recalled the factory at the edge of Lisbon and its secret processes for producing composition cork. "What I remember most about that is fantastic lunches. We'd go near the cork forests where there were these *tavernas* where you would have long, long wine-sotted lunches with beautiful salads and sardines from the sea. Wonderful."

Ginsburg took quiet pride in fostering Blumenthal's management experience, just as he prided himself that Crown International cultivated local talent in the overseas subsidiaries. Ginsburg knew how to work across borders of culture and custom, how to bring along local managers, and how to do business in many countries. Melchor Marsa had provided an important guide to that cosmopolitan management approach. Ginsburg's rise from mail clerk to senior international management reflected an unusual path, the kind Americans mythologized in Horatio Alger stories but that was rare in real life. But Ginsburg did not romanticize his career in the cork industry. He made a respectable CEO-level salary of $60,000 plus an annual bonus of $30,000, for an annual income of $90,000 (over $739,000 in 2017 dollars). When he and Bobbie went into New York for a weekend, they stayed in a suite in the Plaza Hotel, usually the George M. Cohan corner in the Oak Room.

Blumenthal and his wife drove to Mount Freedom to visit the Ginsburgs at home and took occasional dinners with them in the city on weekends. In those situations, Blumenthal always felt himself to be an awkward combination of subordinate and family member.

In all the dinner conversations that the two men shared, Ginsburg almost never talked about the episode during the war. Later Blumenthal recalled vaguely that Ginsburg once came close to revealing the role that Crown Cork played in the intelligence war. "At some point, he mentioned something about, 'Well you know [the government] came to me and I tried to give them some help.'" When Ginsburg sat down with the OSS recruiter in 1942 and explored options for undercover work, he was probably already wary of politics. At the same time, he would want to help. He was of course adamantly opposed to Hitler, as a Jew and as a loyal American.

"Herman Ginsburg is an example of success—great success—in business by people who used management styles that every business school will tell you is a formula for disaster," Blumenthal said later. Ginsburg and McManus "managed in a way that is considered in business schools, particularly today but even then, to be antithetical to good management."

"Business schools will tell you, you have to think about succession, a rewards system, motivation, team building. Bullshit! Ginsburg paid no attention to that. And yet he was very successful. . . . Under certain circumstances, a very hardworking, intelligent, highly focused, motivated guy, willing to work sixteen hours a day, can overcome a lot of impediments that others would normally encounter when they try to manage that way."

That idiosyncratic model, Blumenthal said, contained "the seeds of its own destruction. Such companies do not survive that person who manages to build it up." Blumenthal also gained a larger lesson in keeping with his toughened idealism: "It taught me a lot: that individuals count, and in a way . . . that theme has always come back to me: people count. Who happens to have his hand on the levers of power at a particular moment matters. History is made that way. It can move in either direction depending on who that person is and the moment in time when he or she can wield that power."

Blumenthal recalls being concerned by the rise of plastics, wondering if the boon of postwar technology would turn the bottle-cap business on its head. He asked his boss, "Hey, what about plastic?" Ginsburg brushed it off.

"That will never happen," he said. "It's a question of taste. It would affect the quality of the drink."

* * *

Toward the late 1950s, the McManus sons lost control of the company when a shareholder and Philadelphia box company magnate named John Connelly asserted dominance of the board and forced them to sell their majority interest in Crown Cork. Both sons went on to other careers in business: Charles Jr. served on various corporate boards and raised money for his alma maters, Bard College and the University of Maryland, and philanthropic causes; Walter went into real estate and philanthropy. Crown Cork and Seal entered a new phase.

Ginsburg saw that he would not be able to preserve the international division's independence under the hard-nosed Connelly. In 1961, after nearly forty years with Crown Cork, Ginsburg retired.

Connelly was counting on Ginsburg's protégé Blumenthal to take over the international operation. But Blumenthal too saw how the industry had changed and that his international business career was at a juncture. When the 1960 election lifted John F. Kennedy to the White House, Blumenthal was eager to join the new administration.

"I was desperate to become a New Frontiersman and go to work in Washington. So I quit." Connelly responded angrily, shouting that he wouldn't let Blumenthal go, that he had influence in the White House, that he would get them to retract their offer. Connelly eventually restored Crown Cork and Seal's financial stability and its profitability in a wider range of packaging and containers. (Today Crown Holdings, headquartered in Pennsylvania, remains a world leader in container products.) And Blumenthal applied his international and business experience in the Kennedy White House, where he served as a deputy assistant secretary of state for economic and business affairs. His business experience, directness, and intelligence made him valuable in the Johnson administration. For LBJ he served as chief US negotiator at trade talks in Geneva. After another stint in the private sector, he returned to public service as US treasury secretary for Jimmy Carter.

Blumenthal had little contact with his former mentor apart from holiday cards and the occasional phone call. Ultimately an episode revealed Ginsburg's conflicted feelings about his protégé overshadowing him in public life. At one point a journalist contacted Ginsburg for an article about Blumenthal in *Fortune* magazine. The published article included remarks that stung Blumenthal as "highly unflattering."

"Underneath, Herman was very competitive," Blumenthal realized. In Ginsburg's view, here was a young man whom he had plucked from a corner of academia and who had now risen to great heights. Ginsburg "was jealous and didn't like the fact that I was now a big shot. A much bigger shot than he had ever been."

Blumenthal called his old boss on the phone. Ginsburg denied the quote. They never spoke again. Yet when Herman Ginsburg died in October 1982, his corporate secretary, Ed Buxton, called Blumenthal and informed him.

The treasury secretary gleaned lessons from his former mentor despite the difficult relationship. Blumenthal appreciated that Ginsburg was ahead of his time in adapting management to the norms and culture of each place where Crown worked. "He firmly believed that local people should run the local subsidiaries. He was proud of the fact that he knew how to work with these local managers and that he would accept the ways of doing business in these countries."

During Blumenthal's time in the Johnson administration, US immigration policy underwent a historic shift. In 1965, the government opened the door for three times as many immigrants in the last three decades of the twentieth century compared to the three decades before. "The result," said Blumenthal, "was an unmistakably new kaleidoscopic racial and multiethnic mix that enriched the country's dynamism and vitality." The shift expanded Americans' lifestyle options, and has meant ever larger changes in business management and innovation. "Go to Silicon Valley and you can see that many entrepreneurs are immigrants or children of immigrants. That's a more recent development.... That didn't exist in the days when I worked with Ginsburg. That's a big change."

* * *

At home Ginsburg rarely talked about his work, but later he did tell his niece and nephew about his start in the cork business as a teenager. And he told Connie, when she asked about his travels, that he had a friend in Portugal. In retirement Herman and Bobbie moved into an apartment in lower Manhattan on East End Avenue, with a striking view of the East River.

The name Ginsburg was often on the lips of Charles McManus Jr. in later years, according to his son David. Long after he left Crown Cork and the cork industry, Charles Jr. talked about Ginsburg and the others who had made a difference: Melchor Marsa, Antoine Leenaards, Ernesto Mas, Josep Pla. David also remembered visiting his father at the Highlandtown factory as a young boy. He got to sit with the engineer as the man managed the switches for the freight cars. They watched the trains come in from the port, felt them chugging by on the rails. The cork came in from Portugal and other places in a steady stream.

As the McManus brothers moved into other ventures involving bottling, real estate, and transport, Crown Cork grew even bigger in the container business, with another generation of workers.

Chapter Ten

# MAKING IT IN AMERICA

· DiCara: 1945 on ·

> There are three trees in the world whose bark yields that which is of more real value to man than all the jewels and precious stones ever dug from the earth. The cuticle of a South American tree yields the liquid which is *caoutchouc*, or india rubber; the bark of the Peruvian cinchona bestows quinine, and the bark of a species of live oak supplies the world with cork.
>
> WALTER A. RILEY, "The History and Use of Corks and Other Stoppers," 1906

In the summer of 1945, Frank DiCara's days in the army encampment in the Philippines grew more anxious and steamy. He was issued quinine and Atabrine to prevent malaria, but he and others came down with the shaking fever anyway. He watched another GI wrapped in five or six blankets, in the sweltering tropics, trembling so hard from chills that he nearly shook them off. More than half of GIs who served in the Pacific reported having malaria at some point. Others suffered differently.

"Some of the guys went a little bit wacky," Frank said. "They couldn't cope with being over there, being away from home. They couldn't cope."

DiCara volunteered for night patrol duty because otherwise he'd be lying awake, obsessing. Across the ravine from camp he saw nothing but sand dunes and the headhunters of his imagination. In bed he thought about how the headhunters would slip across the gully, enter his tent, and kill him. Other times he pondered how the Japanese dug tunnels with entrances that they covered with leaves. They'd crawl through the tunnels at night, climb out, and stab him in his cot if he didn't stay awake and alert. So he didn't sleep much.

On patrol duty, he and Donald Dubb, a grunt from New York, debated which was worse: What would happen when they boarded the boats to attack Japan or the dangers that lurked on the next hillside? They also talked about the survival struggle that the Filipinos around them faced.

Frank had known poverty growing up in Highlandtown, but the desperation of families he saw on Mindanao shocked him: the lack of food, their only water coming from a communal spigot. The Philippines had fared terribly under US military rule for decades after the Spanish-American War, with a transitional democracy since 1935. Frank saw how the Second World War had brought famine and destroyed homes and farms. With liberation from the Japanese came new struggles. Irrigation systems lay in ruins, the currency was nearly worthless, and families were hostages to profiteers. The government tried to limit the exploitation by prescribing maximum prices for crucial goods, but nobody enforced those caps. People scavenged and foraged at the edges of GI camps.

One day after seeing their suffering, Frank went to the mess hall, grabbed a large sack of rice, and hoisted it on his shoulder, not exactly on the sly. He went into town and gave the rice to girls staying next door to the place he and his buddy rented.

The GIs' relationships with the locals were complicated and confusing for a teenage soldier. There was a lot that Frank didn't understand in the dynamics with them and the Japanese prisoners of war. Frank supervised Japanese prisoners who did painting and maintenance tasks in the camp, and started thinking that maybe they weren't so different from him. A Japanese captain who painted the mess hall spoke English so well—better than Frank—that Frank couldn't believe he was Japanese. Before the war, the captain said, he had studied at the University of California. Frank had trouble squaring his understanding of that man with the utter strangeness of kamikaze pilots dive-bombing Okinawa, killing themselves for a cause that everyone saw was hopeless. That was how they were raised, he supposed: *Your life means nothing, except for the empire*.

Frank hired one of the Japanese prisoners as a valet to keep his tent clean. The man quickly found Frank's bottle of Calvert gin and polished it off. Frank found him drunk, nearly passed out in the tent. Furious, Frank pulled out his pistol to scare him.

"No, Joe!" the man cried. "Please, no." Frank felt a jolt in the hand that held the pistol. The memory stayed with him.

Another morning, Frank told the valet to take a chicken to the girls' family in town. When Frank asked about it later, one of the girls told him that the chicken never arrived. The valet had eaten it. Frank exploded again. "The next time I'm really gonna shoot you!" he shouted.

Equipment came and went through the camp. The B-26 pilots continued their runs. Far off, generals were masterminding the details of Operation Downfall and the invasion of Japan. Frank's tank battalion would be joined by other parts of the Sixth US Army. The army also crafted a deception plan labeled Operation Pastel, designed to convince the Japanese that the Allies had decided to blockade the islands ahead of a bombardment. The invasion would begin in November.

On Mindanao one sweltering day in August, when the boredom and anxiety had reached a simmering pitch, Frank was in his tent, not thinking much of anything. He listened to the radio as background noise. Suddenly a voice came on with an astonishing announcement: a second atomic bomb had exploded on Nagasaki. Japan had surrendered.

The war was over.

The encampment went wild. That night they heard the low, loud drone of foghorns out on the bay, navy ships sounding off in celebration. The drinks flowed. And stumbling back to their tents, some felt a strange lightness.

Yet in the days after, an undercurrent of fear remained. What came next? Frank knew there were still many Japanese hunkered in caves across the islands who wouldn't know or believe that the war was over.

Bureaucracy survived the war too. The days contained as much frustration as relief. When could they go home? Nobody knew.

When their orders finally came, Frank boarded a tub called the USS *Dash and Wave*. The name sounded like a joke, and it clanked away from the dock, barely stirring up a wake. The army didn't share Frank's sense of urgency about getting home.

Out at sea, a ninety-mile-an-hour gale rocked the *Dash*. The ship was tossed like a toy. A man couldn't stay upright on deck. Down in the hold, Frank heard the deep rumbling and groans of wrenching metal, as though the boat were going to crack.

"Frankie," his buddy from Tennessee said, "I knew we were never gonna make it home. This ship's gonna break in half!" After all they had made it through—crossing the sea, dodging bullets and disease in malarial jungles, and fevered dreams—Frank suddenly felt the same certainty as his buddy: they were never going to get home. The boat pitched in its last throes. By morning, the sea had calmed.

* * *

In the months leading back to civilian life, millions of soldiers prepared to reenter the American workforce. The huge number of servicemen seeking jobs, and the reduction of defense-related work, raised concerns. In a 1944 Gallup poll, nearly half of all Americans surveyed expected a large surge of unemployment—up to 34 percent—after the war ended. The government itself forecasted 25 percent unemployment. Frank got training that the officers said would help him reenter the work world. He became skilled at leatherwork and jewelry making and got a certificate in refrigeration and air-conditioning engineering and repair. He didn't know what economy he was coming back to, but these skills might give him an advantage. Plus he had factory experience and was a veteran. He felt confident.

The terms of DiCara's draft notice said "duration of war plus six months." In the spring of 1946, he returned to Baltimore and found tremendous changes. He was completely changed now himself.

Life at his mother's house was familiar and yet very foreign.

Neighbors asked about his plans now that he was home. They didn't ask or seem to want to hear about his experiences during the war. It was safer to talk about his job training and where he was looking into new opportunities.

The factory where he had worked on the bombers seemed a good place to start. When DiCara went to the personnel office at Crown Cork and Seal, no record of him could be found. He was told that he had not worked for Crown and to go to Martin. So he did.

Martin's aviation business was still popular, but it was quickly winding down wartime operations and was growing famous for its layoffs. A joke around town was that Glenn Martin had built his first airplane in a barn with his mother holding a lantern for him to work by and that as soon as he could afford to buy a hook to hold the lantern, he laid off Mrs. Martin.

At the Martin personnel office, DiCara got more deflection. They took a few minutes to look for his records in their personnel files and then they told him they had no record of him. He must have worked for Crown Cork and Seal. After all, they asked, what name was on the paychecks he received? He would have to go back to Crown Cork.

All this was deeply discouraging. One day while he was out hunting for work, Frank had an experience that captured the atmosphere of that time. He boarded a streetcar and found a seat near a group that seemed happy. They were talking excitedly, laughing about how business was good. But not as good as it had been during the war, someone said. The guy sitting closest to Frank laughed and said, "I wish the war would have lasted a few more years! I was making a lot of money then."

Frank didn't say anything. His status as a veteran wasn't obvious—by then he had stopped wearing his uniform. But the more he thought about it on the ride home, the more the guy's words made Frank's blood boil. Frank realized: *That's the feeling of the American people. This guy didn't know the experience of lying in mud or in a foxhole filled with water. He had never felt tropical heat or frostbite. He just wished the war would've gone on for two more years. Isn't that a wonderful thing to say?*

Feelings of alienation were common among returning veterans. As Thomas Childers notes in *Soldier from the War Returning*, polls in 1947 found that nearly half of US veterans felt they were worse off than before the war and that they had lost the best years of their lives; around one-third felt estranged from civilian life. One veteran in five felt "completely hostile" to civilians.

Frank went back to Crown Cork and braced himself for one more effort. He told the man there was a mix-up. After all, his old pay stubs all said "Crown Cork and Seal." So he must have worked for them. Still the man kept shaking his head. They didn't have a job for him. Frank turned around again and left.

After that, as though in a daze, he made the rounds to other factories and shops. He was nineteen and he didn't have a high school degree. His brother Joe, who had been awarded a Bronze Star for capturing a German machine gun nest, was back home selling produce from an arabber cart, the century-old tradition of Baltimore street vendors. New companies were coming up, but they wanted different skills, not necessarily in refrigeration. Weeks turned into months. Then nearly a year.

One day on a street in Highlandtown, a man walked up asking for directions. Frank recognized the man as his old boss, Mr. Storrs, the general foreman at the Marauder wing plant. He didn't recognize Frank until Frank said, "My God, Mr. Storrs! It's good to see you."

The man was stunned. "What are you doing, Frank?"

Frank looked different; he was no longer a green teenager. He explained how he had returned from the war and was looking for work. He had an application in at Baltimore Gas and Electric and waited to hear from them. The power company was hiring as the city prepared to replace the last seventeen thousand gas lamps with electric streetlights.

"Come down Monday morning," the man told Frank, "and we'll get you a job."

So DiCara went for a third time to Crown Cork's office to apply for a job. This time Mr. Storrs made it happen. A few days later, DiCara clocked in for his job on the lowest rung of the ladder: a spot on the yard gang, slinging cork and shoveling coal. In the open yard beneath the huge silo where the cork came in, the yard gang piled up the big bales of cork. Each one weighed more than one hundred pounds, sometimes closer to two hundred. The barges from Portugal stood in the harbor just out of view, but their loads came by rail and weighed down Frank's days. A full shift with the yard gang was as dirty and bruising as a longshoreman's.

At the end of each shift, Frank dragged himself home and took a second shower for the day, washing off all the sweat, dirt, and cork grime.

The yard gang introduced him to the full range of characters in the bottle-cap business. They included the bookmakers who managed the betting. Some of the guys tracked the newspapers, and the space where the US Treasury announced its daily balance statement. Bookies used that number from the daily paper—usually the last three or five digits—for calling the winners on that day's "Treasury balance" bet. Some bookies handled horse races too. One guy, sometimes a foreman on the yard gang, was what they called a "juice man"—the lender of last resort, a loan shark. Many black workers borrowed from him, since they didn't have access to bank loans. They had to repay him six to five, or 20 percent interest. Every payday a line formed in front of the juice man for repayments.

As hard as the yard gang work was, Frank was relieved to have a job. He was starting to feel like maybe, in this strange world, he had a place.

Frank has a memory of seeing the old man, McManus Sr., during a visit to the factory floor. "I think he was a compassionate guy," Frank later said of McManus. The sons were accessible. Charles Jr. and Walter McManus would come down from the office and mix with the workers at the Christmas party, when everyone and their families came, but also during the work week. Frank could go over and talk with them like they were regular people.

Frank had come a lot further than just across the factory floor. From seeing the glow of that September 1940 factory fire out his bedroom window on Pratt Street, he had made his way to the company that dominated Highlandtown. And with his job on the yard gang, he was not yet done with his incarnations in the cork industry. He was open to persistence and opportunity. Like cork itself, one could say.

* * *

The container industry faced turmoil after the end of the war. Armstrong Cork reported that the scarcity of materials it had experienced through the war continued for many months after. Besides shortages of rubber and cork, inputs like zinc oxide, synthetic resins, and paints were also scarce. Manufacturers experienced a severe shortage of linseed oil throughout 1945. Every day, the corporate chemists got requests from management about substitutes, recalled one flooring chemist. "What can you substitute for this? What can you substitute for that?"

New synthetics were key. For building and flooring materials, Armstrong introduced a material called Accotex, which initially cost more than cork but lasted up to six times longer. There was also a shortage of tinplate for metal caps. Industry's appetite for materials was shifting. Plastics were cheap and adaptable. Soon after the war ended, thousands of people flocked to the first National Plastics Exposition in New York to see the new materials and products that were possible: fishing line as strong as steel; lightweight, resilient suitcases; and clear containers that sealed food better than ever.

Domestically grown cork and composition cork could not compete. Woodbridge Metcalf kept with the cork-growing program in California through 1947, by which point the state had produced a half million cork oak trees, at a cost to McManus of 12.5 cents each. Metcalf's forest extension network distributed them through state rangers and farm advisors from San Diego north to the Oregon line. They stripped around five tons of cork from mature trees. Metcalf noted with some pride that chemists at Armstrong Cork and Crown Cork found California cork to be "practically as good as any European cork." But cork quality from a tree goes up with every harvest, with many trees in Portugal and Spain continuing to produce for a century. For lack of previous harvests, those in California didn't measure up.

"I never did think that the project was a feasible thing economically in California," Metcalf admitted late in life. With the high costs of labor and with land that could be more profitably farmed in other crops, growing cork made sense only in a crisis. "The United States has always used 50 to 60 percent of the world's supply of cork," he continued. "They can, with the cheap labor that they have in Spain and Portugal, strip the cork, bring it down to the ships, and land it at the U.S. Atlantic coast ports . . . at the time we were working on it, for about five cents a pound."

Still, the experiment of McManus's project showed Americans where they could grow cork if economic conditions changed to the point where it became competitive with other crops. "A lot of the trees still grow on the [University of California] Davis campus around the quadrangle," Metcalf added. "Beautiful trees."

The planting guidelines and climate maps for different cork varieties, published in academic journals and glossy popular magazines, piled up in library stacks. Giles Cooke continued to promote cork plantations, producing articles for another decade. In 1961 he published his book *Cork and the Cork Tree*, which he dedicated to his wife, Anna, and daughter, Ann Dorsey. Along with his research and an account of the McManus Cork Project, the book included the earliest observations on cork plantings in America, by naturalist John Bartram in 1765, Thomas Jefferson's correspondence, and the latest instructions for planting and care of young cork oaks. "Certainly a material with such an impressive record over so many years deserves more than casual attention," Cooke wrote. "The more we know about cork the greater will be the benefits derived from it by mankind."

Yet Cooke realized that at least for another generation in the era of plastics, modernity had left cork behind. The industry where Frank DiCara found work was already shifting for a second time, into new kinds of containers and bottle caps.

* * *

With a steady job at Crown Cork, Frank found his life was coming into focus. He was twenty-one years old and in love with his childhood sweetheart. Irma had finished high school and had a stenography job. Her father, Anthony Castagnera, worked in construction. Like Frank's father, Mr. Castagnera had seen immigrant workers miss out on opportunities that other Baltimoreans got. Castagnera always said that the game was rigged. "What makes the rich rich is that they're stealing from the poor," he would say over dinner. "They take your money. They make big money and pay you nothing."

Frank's generation of veterans, returning from the war after passage of the 1944 GI Bill, gained a foothold into the middle class. The GI Bill provided veterans with money for tuition to attend college or university, as well as low-cost mortgages. Any veteran who had served on active duty for at least six months and did not have a dishonorable discharge could apply. In the eleven years following the war, nearly half of the 16 million World War II veterans received training through the program.

Frank's job at Crown Cork opened a door to the life that he had always wanted, just as the business had taken in Melchor Marsa and Herman Ginsburg before him. With that security, Frank proposed to Irma, saying, "Could I be the man for you?"

Irma came from a big family with aspirations of financial security. She considered Frank's proposal and replied seriously: "You're a good worker. I know you're going to be a good provider."

Frank's mother, happy with the match, told her son that the place to go was S&S Katz. Go and buy a ring, she told him. So Frank walked over to Eastern Avenue and picked out a ring for Irma.

They set a date for their wedding at Our Lady of Pompei. Irma's father refused to take part in the ceremony or give her away at the altar because, he said, the Catholic Church was part of the corrupt system. So Irma asked her grandfather if he would escort her down the aisle instead. When her father heard that, he changed his mind: if anybody was to walk his daughter down the aisle, it would be him.

After the ceremony, the families celebrated at Gallagher's Hall on Erdman Avenue with an orchestra, open bar, and all the food their family, friends, and Frank's Crown coworkers could eat. All on a $300 budget.

Frank and Irma couldn't afford a honeymoon, but the newlyweds got the next best thing: three months free in the apartment of a friend who also worked at Crown Cork and was going away with his wife for three months. After that honeymoon in Highlandtown, the newlyweds moved into an apartment building that Irma's grandfather had built. They paid $35 a month in rent.

That was the first part of what Frank called his destiny. In his early years with Crown Cork, he heard about how McManus Sr. had brought some of the New Process Company crew with him when he took over the company back in the 1920s. "McManus," Frank said later, "he was a compassionate guy."

The second part of Frank's destiny came later, after three or four years slogging away on the yard gang, piling cork and loading trucks. This piece of his destiny found him walking upstairs at Crown Cork to get something. That's when he ran into a superintendent named Tom Healy. Healy asked, "What are you doing?"

Frank explained that he had come back from the war and was working there in the yard gang. Healy said, "Monday morning, you come and see a guy named Howard Cantwell. He's starting a quality-control unit. Tell him I said to put you to work. I'll transfer you."

That started DiCara's rise into management and a four-decade career at Crown. Destiny put him in front of Tom Healy, Frank said. But Frank made his own opportunity. He worked his way up through assembly-line quality-control inspection to the litho department, which printed the labels for cans and bottles, and on to wholesale customer relations. He worked with Giles Cooke on inspection of cork and temperature checks. Frank traveled to troubleshoot production problems and helped to start new factories from Alabama north to Canada. The container industry continued to change, and Frank made his way amid those changes.

His family still reminds him of the old times, and what they overcame. His nephew says, "Uncle Frank, remember when you four were all in the service and they came and took the shortwave radio out of the house?"

Frank says, "Yeah, I remember that." But many people didn't remember those episodes, and many families that had been labeled enemy aliens didn't talk about it. The experience of Italian Americans during World War II was a chapter that history books overlooked until the 1990s. Frank's education, from that alienation at home to war and poverty in the Philippines, was hard to convey to others. "How can I instill that I have seen death, that I've seen poverty, that I've seen sadness, that I've seen people that, if you have any compassion, it would break your heart?" he asks. "How do I relate that to someone who didn't see it?"

DiCara rose to foreman and then general foreman. And then a higher management position came open, dealing with customers and managing a sales team. Two supervisors tapped him for that job. He questioned whether he could fill the shoes of the previous manager, who was well liked. Two months after he took the manager job, the same two supervisors came to see him, and he was sure he had failed his probation period. They said, "Frank, we have to apologize. We thought that nobody could do the job like your predecessor. But you've gone beyond that. You are doing better than what he did."

"I have used this with my grandchildren," he says of the promotion story and his doubts. "Don't ever think you can't do the job that somebody else did. Your mind is just as good and you've got to use it."

# Epilogue

# TREASURY BALANCE

The uses to which corkwood may be put are unlimited. . . . A wonderful material truly, and of interest, so full that it seems I have failed to do it justice in my humble endeavor to describe the Quercus Suber of Linnaeus—Cork.

WILLIAM BOYD, "Cork," in *Granta* 34, 1992

He assembled a collection of really, really ugly plastic chairs. To him they were so interesting because they showed how a material and its possibilities can bring out the best and the worst in people.

PAOLA ANTONELLI, quoted in *Plastic: A Toxic Love Story* by Susan Freinkel

Seventy-seven years after the spectacular Crown Cork fire, Frank DiCara lives within sight of the Baltimore skyline and not far from the old factory. At his home he holds up a bottle cap to inspect it—a red-and-gold item off a ginger beer. He touches the plastic inner liner with the critical assessment of a connoisseur.

"If you open a bottle at a bar, you'd never imagine what that went through—all the checking the quality, the color. You'd think it was just nothing. Bump! Bing." He snaps a finger.

DiCara may have been a wisp of a boy, but at ninety, he's sturdy as a stone wall, with streaks of white combed over the top of his head. His pink face registers everything. His character is concentrated in his voice, which sounds like cement poured through a coarse sieve: a fibrous Baltimore baritone with the grit of ages.

Frank made a career from bottle caps at Crown Cork, and from that, a better life for his family. His children enjoyed the fruits of education and financial security. Life was not all sweetness and light, he admits.

He and Irma were married sixty-five years when she died in 2012. Photos of his daughters and his grandchildren line his living room.

Today the Highlandtown factory sits idled. In its maze of buildings, a few artist studios give the appearance of life, but manufacturing hasn't provided jobs in Baltimore for decades. The neighborhood now is predominantly families from Mexico, Puerto Rico, and El Salvador. Inequality in the city is worse than what it was in 1945. Highlandtown's median household income in 2010 was $28,813, with nearly one in four residents living below the poverty line. Nearly one in six housing units stood vacant.

* * *

A four-hour drive south of Baltimore, down the Eastern Shore, I visited an old cork oak. The tree stood fifteen feet from the road, about a half-mile south of Capeville, Virginia. It was nearly sixty feet tall and a bit wider than that at its crown. I walked up and laid my hand on its massive trunk. The irregular bark had a startling spongy feel. It probably hadn't been peeled in half a century. Only a few limbs still sprouted green leaves; the tree was slowly dying. It turned out to be older than the trees planted by the McManus Cork Project, dating back to around 1907—about the time Melchor Marsa and Frank DiCara's parents arrived in America. I circled around the tree, patted its gnarly skin, paid respect. It was a survivor.

The US cork industry endures too, although much changed. It now accounts for not quite 22 percent of the world market, compared with 50 percent at its peak. Plastic screw-tops have mostly replaced cork wine stoppers. US sales of cork and cork products total $150 million annually. (Cork stoppers account for two-thirds of that total; composition cork, flooring, and other products make up the other third.)

So cork's place in the landscape has changed. In cities on the East Coast, cork factories fell into disrepair, were paved over or converted to condominiums (in Pittsburgh), a hotel (in Lancaster), and a space for artists' galleries (in Baltimore). The same energy that shaped the factories in the first place has reshaped them, and it animates the creation of electronics plants, new stores, and auto dealerships.

You can almost imagine a shadow landscape—what might have been if composition cork had remained the state-of-the-art sealant. A landscape with cork savannahs across the Southwest, local businesses with semiskilled jobs for harvesting and processing, and cork-manufacturing facilities where plastics companies now stand.

Few traces remain of the millions of seedlings planted by the McManus Cork Project. The ten acres planted by a dentist in the Carolinas—gone. Acres of cork in California, Arizona, Mississippi, Maryland, and elsewhere in between fell to neglect, were cleared and replanted with other crops. But individual reminders still exist. In Flat Creek, South Carolina, a crossroads sixty miles northeast of Columbia, a farmer named Clyde Pittman planted cork seedlings near his home in the 1940s. On his farm in the state's Sandhills region, Pittman already had acres of pine alongside native hickory, sweet gum, and other hardwoods. But he was especially excited, his daughter Glenda recalls, to add cork trees to the grove. He watered the seedlings, tracked their growth, and for years got a thrill from showing visitors the small trees with the strange bark.

"It was a conversation piece," Glenda said. "Most folks had never seen one before." One of his cork trees still grows near the family homestead. Glenda has photos of her grandchildren seated in its branches, goofing and smiling.

Two hours in the opposite direction from Columbia, in Brunson, South Carolina, lives Sarah Harrison Corbin, who in 1944 was a 4-H Club member and planted cork seedlings. She was about ten years old. "Both came up," she told me of the seedlings she and her brother planted. "One died but the other one grew to be a big tree." She laughed and said, "We always said *his* died and mine lived." The trees were near their parents' home, amid the family's orchard of peach, apple, and plum trees.

As the years passed her memory became less clear, and it felt almost like the cork trees had never happened. "I would tell people I planted a cork tree and they told me I was crazy. 'That wasn't a cork tree,' they said. 'It was an oak.'" She was relieved to learn that her memory was correct.

Then she asked, "So what difference did it make? What did it mean?"

I shifted the phone to my other hand and replied that her tree and thousands of others showed that cork could grow here if needed. Those words sounded lame as I replayed them later in my head. Her question deserved more. The truth was, after the spotlight of the cork-growing campaign, with the attention of thousands of young people and foresters, cork fell out of the public eye. What *did* it mean? Like so many times in history, humanity focused its gaze intently on a fellow creature, altering the course of that creature's fate on earth. And then we moved on.

Like forks of an oak, the future sprouted a flush of branches and possible outcomes—some proved vital and grew, and others were pruned and sealed off—shadow futures. The cork project was one of those shadow futures that was pruned away.

A branch that proved sturdier, in one sense, was the drive shared by the people behind the Cork Project. They believed that we could plan for future supply on a large scale and innovate, bringing forth new energies. In the 1960s Michael Blumenthal, Ginsburg's protégé, had a front-row seat at the White House to America's shift on immigration, and a more open approach. That open-door policy spurred American enterprise through the second half of that century and shaped the landscape of our own century.

Through the branches of the future that materialized, come glimpses of the scenes from over a century ago, in the burnt, discarded buoy that John T. Smith found among the ashes of his firebox by the East River docks and his accidental discovery there of composition cork. The young McManus testing sealants in his kitchen sink in Manhattan. The waves of ships arriving in New York Harbor, bringing immigrants down their gangways, some bound for work in the industry. All the shades of faces and their expressions as they scanned new surroundings. A twenty-three-year-old Melchor from Barcelona seeing Liberty's statue. A five-year-old Herman with his parents arriving from Lithuania. A young couple touching shore from Palermo, having met on the ship, passing through the Immigration agents' disinfecting shower.

The story of cork in wartime intrigue was buried for decades, partly by the technological churn that sidelined cork itself and partly by the national security machinery that left it hidden for decades, locked in classified government files. Charles McManus Sr. was not inclined to write down his story, nor was Melchor Marsa. In an oral history conducted later, forester Woodbridge Metcalf was alert to the McManus Cork Project's drama, but he was above all a practical person, and the tale no longer seemed to have practical value.

The story of the McManus Cork Project found me while I was pursuing other topics. I became intrigued by what seemed a big, quirky gamble. The range of characters fascinated me. The few articles written for the general public during cork's heyday were florid publicity pieces. But a few conversations with McManus Jr. and Frank DiCara brought the wartime era alive. Trips to the National Archives turned up declassified memos and the intelligence roles played by Melchor Marsa and Herman Ginsburg. Conversations with others, especially Gloria Marsa Meckel in Mexico, breathed life into the narrative. So did the memories of Treasury Secretary Blumenthal.

The wartime pressures of the story echoed recent episodes when a natural resource suddenly became scarce and seized with great importance. The item could be water, or petroleum. Academics have studied more closely the ways that conflict affects natural resources. Another case: rare earth materials are crucial ingredients in the powerful magnets used in modern electronics, everything from computers to electric motors. Dubbed "rare earths," they're becoming increasingly scarce worldwide. Scarcest of all are the heavy rare earths. (Anything called "heavy rare earth" just *sounds* scarce and valuable.) Experts predict these elements will be exhausted within twenty years. The main source is clay from the soil in southern China. Despite a search for substitutes, the United States still relies heavily on imports. Finding domestic alternatives has become a matter of national security, some say, since the nation's defense, economy, and lifeblood depend on those mineral elements for electronics. Corporations and public agencies have found almost 11 million tons of rare earth resources in coal deposits across nine states and are investing in new technologies to extract rare earths from the coal.

In those cases, the effect of scarcity and conflict can be like flipping a switch: suddenly the people who work with the material in their day-to-day work get snared by powerful forces. The change is like the one faced by a fisherman who steps in a coiled line, when the line gets pulled taut by a fleeing marlin.

The cork industry drew its workforce from many marginalized neighborhoods and attracted a number of outsiders. The McManuses, Marsas, DiCaras, and Ginsburgs may or may not reflect the industry at large. Yet their story shows people's resilience and inventiveness in difficult times.

What does that look like today? What pressures does an obscure resource and scarcity create for people who rely on it?

In the twenty-first century, cork forests in Spain, Portugal, and Morocco have become a cause for biodiversity preservation. The forests shelter Mediterranean wildlife: lynx, pigs, and red deer, among others.

But with market demand for cork down—and cork stoppers for wine losing ground to screw caps—other land uses have become more profitable. A World Wildlife Fund report in 2006 forecast that in time, only 5 percent of all wine bottles would be stoppered by cork. So the cork forests are left untended, undergrowth becomes overgrown, and the risk of wildfire increases. Landowners have left their cork groves in record numbers, moving to the city. Their absence in the countryside has made the risk of wildfire even greater, as recent fires have shown. In Portugal, more than sixty thousand people still make a living in the cork industry. Shifts in the use of cork continue to destabilize the environment and economy and could cost thousands their livelihoods.

One lesson is simple: understand your interests. Each person faced their own war, whether it was a secrets war in Portugal or the confusing home-front war of camouflaged factories and enemy alien restrictions, or the return from the frontlines to a changed world. Charles McManus was propelled by his curiosity, problem-solving ability, and intuition. The industry suited his interests and rewarded him. Melchor Marsa came to the industry by chance; through a caprice of geography and geopolitics, he became nearly consumed by it during the war. Frank DiCara discovered his talents on the fly and followed them to a better life.

# ACKNOWLEDGMENTS

There is a lot that we will never know about America's cork industry and its involvement in World War II, partly because the industry was eclipsed by a new idea of modernity after the war. Furthermore the records of agencies, especially the Office of Strategic Services, or OSS, and its successor, the Central Intelligence Agency, are shrouded in secrecy. In finding this story, I owe debts to many people. My detailed obligations appear in the essay on sources. More generally, for suggestions and insights, I am indebted to Thomas A. Guglielmo, Clint Richmond, and Dr. Douglas L. Wheeler, who generously shared chapters of his work-in-progress. Dr. Amélia Branco, an economic historian at the University of Lisbon, shared her insights and references on the *montado* ecosystem and the cork industry in Portugal during the Estado Novo (1933–1974). And I am grateful to the late Aline, Countess of Romanones, for her personal memories of working with the OSS. They all answered rounds of questions with patience and encouragement.

Librarians have been a great resource all along the way, including: Susan Rigg at the Swem Library of the College of William and Mary, Matt Shirko at the Baltimore Museum of Industry, and the librarians at the Bancroft Library of the University of California at Berkeley, the reference staff at the National Archives in College Park, and the librarians at the Enoch Pratt Free Library in Baltimore. At the Library of Congress, I'm indebted to William Elsbury and Thomas Mann, who has since retired from the library but remains a master at finding hidden sources of information.

Researchers Vicki Rushworth and Maria Brandt made valuable finds among the materials in the Donald Downes and Woodbridge Metcalf collections, respectively. Catarina Fernandes Martins provided translation support in several Lisbon archives, and hosts Luís Oliveira and Antónia in Évora acted as interpreters with cork processors at Azaruja. Thanks also to editor Peggy Goldstein, who did a remarkable job of making clear prose from passages that did not start that way, and Michael Baker for his copyediting.

For support during the research and writing, I am deeply grateful to the Cork Institute of America, especially Jerry Manton and Art Dodge, and to Amorim Cork, especially Carlos de Jesus.

While writing this book I enjoyed the nurturing environment of a residency at the Virginia Center for the Creative Arts, where fellow writers and artists listened to my obsession and provided remarkable feedback and encouragement.

Albert LaFarge, my agent, has been a tireless champion of this project. Thanks also to Elizabeth Demers and Matt McAdam, my editors at Johns Hopkins University Press, for their perceptive guidance. Matthew Algeo, Deborah Furlan Taylor, and Andrew Wingfield all provided smart and necessary suggestions on early drafts. I'm deeply grateful to my writing group for their insightful feedback, questions, and encouragement: Angela Chuang, David Ebenbach, Melanie McCabe, and Emily Mitchell, brilliant writers all.

I'm most grateful to my family on what has been a longer journey than expected. My parents were always interested, and I miss their questions and encouragement. I did not work fast enough for them to read this, but their support made this a better book. Above all, I thank Lisa Smith, who has supported me along so many journeys, from drives down Maryland's Eastern Shore searching for cork trees, to travels through Portugal. She enriches my life with her priceless enthusiasm, keen eye, humor, encouragement, insight, and love.

# ESSAY ON SOURCES

## PROLOGUE. THE BLAZE

The 1940 factory fire at Crown Cork and Seal was a disaster that nobody who witnessed it would forget; of the families featured in this story, the two who were there recall it vividly. My account of the fire is based on my interviews with Charles E. McManus Jr. and Frank DiCara, complemented by news accounts, especially from the *Baltimore Sun*. The overview of cork's history comes from several sources and owes much to Arthur Louis Faubel's *Cork and the American Cork Industry* (New York: Cork Institute of America, 1938); Giles B. Cooke's *Cork and the Cork Tree* (New York: Pergamon Press, 1961); and *The Art of Cork* (Porto, Portugal: Corticeira Amorim, 2014), produced by Amorim, which contains an illustrated account of cork's history and properties.

The account of the cork cargo fire in New York Harbor comes from the *New York Times*, "Part of Cork Cargo Burns," Sept. 30, 1941. For the narrative of the Marsas throughout the book, I owe a great debt to Gloria Marsa Meckel. My interviews with her were supplemented by conversations with Melchor Marsa III. The discussion of the McManus Cork Project here and in later chapters owes most to the careful documentation of Giles B. Cooke, whose papers are maintained at the Swem Library at the College of William and Mary, and to Sam Sheldon's essay "The Cork Forest Industry in the United States," *Maryland Historical Magazine* 99(1) (Spring 2004): 95–110.

## CHAPTER ONE. MCMANUS PEELS THE APPLE

The narrative of the McManus family and the McManus Cork Project during the war started with my unexpected finding of the project's existence during research on another history; it gathered momentum with my conversations with Charles McManus Jr. in January and March 2006. In addition to my interviews with him, this chapter's description of the New York World's Fair of 1939–40 comes from several accounts, including "Exhibits: A World of Wonders" in the *New York Times* (May 5, 1940), the *WPA Guide to New York City*, and Forrest H. Taylor's memoir, "The World of Tomorrow—1939 New York World's Fair" (Waynesburg University's Teaching with Primary Sources program, http://tps.waynesburg.edu/documents/901-the-world-of-tomorrow-fh-taylor-memoir/file).

My description of Baltimore in 1939 and 1940 comes mainly from the *WPA Guide to*

*Maryland: The Old-Line State* (New York: Oxford University Press, 1940). McManus Sr.'s working method comes in part from his profile in *The National Cyclopaedia of American Biography* (New York: J. T. White, 1984).

The assessment of American attitudes to enterprise and the government's role in society is informed by Jodie T. Allen's Pew-supported study, "How a Different America Responded to the Great Depression" (Pew Research Center, Dec. 14, 2010, http://www.pewresearch.org/2010/12/14/how-a-different-america-responded-to-the-great-depression/).

For more on the explanation of Crown Cork and Seal's origins and the discovery of composition cork, see the book by William Painter's son, Orrin Chalfont Painter, *William Painter and His Father, Dr. Edward Painter: Sketches and Reminiscences* (Baltimore: Arundel Press, 1914), and Pearl Edwin Thomas's *Cork Insulation: A Complete Illustrated Textbook of Cork Insulation—The Origin of Cork and History of Its Use for Insulation* (Chicago: Nickerson & Collins Co., 1928). The section on the effect of Prohibition on the industry and McManus's early career and marriage to Eva Olt comes from my interviews with Charles McManus Jr. The description of McManus Jr.'s early years comes from those interviews, as well as from phone interviews with his daughter, Eva M. Edmonds. The vivid descriptions of the cork harvesting process I owe to Charles McManus Jr.

Two principal resources for learning more about Armstrong Cork during this period are William A. Mehler Jr.'s *Let the Buyer Have Faith: The Armstrong Story* (Lancaster, PA: Armstrong World Industries, 1987) and the biography *H. W. Prentis, Jr.* (Lancaster, PA: Armstrong Cork Company, 1961). The speeches by the indefatigable Prentis appeared at length in newspapers, especially the *New York Times*: "Madden Is Accused by Prentis of 'Tricks'" (Jan. 26, 1940), "Chamber Weighs Our Part in Peace" (May 2, 1940), "Manufacturers to Meet: Chicago Conference Will Make Defense Cooperation" (May 25, 1940), "Defense 'Blacklist' Charged to Unions" (Oct. 18, 1940), and "Prentis Condemns Draft of Wealth" (Dec. 13, 1940).

On the industrial lead-up to the war, an excellent resource is A. J. Baime's *The Arsenal of Democracy: FDR, Detroit, and an Epic Quest to Arm an America at War* (Boston: Houghton Mifflin, 2014). Robert J. Brugger's *Maryland, A Middle Temperament: 1634–1980* (Baltimore: Johns Hopkins University Press, 1996) informed the Maryland portion, including the observation about Crown Cork's scale-up from bottle caps to nine-thousand-pound gear rings, on p. 537. The Federal Writers Project guide, *Maryland: A Guide to the Old-Line State* (cited above) again proved helpful for getting a flavor of Baltimore at the time. And for the texture of national life in years before that, Bill Bryson's *One Summer: 1927* (New York: Doubleday, 2013) provided evocative details.

For other aspects about the cork industry and the prewar setting and industry, I recommend Giles B. Cooke's *Cork and the Cork Tree* (cited above) and again, Arthur Louis Faubel's *Cork and the American Cork Industry*, as well as V. A. Ryan's *Geographic and Economic Aspects of the Cork Oak* (Baltimore: Crown Cork & Seal Co., 1948) and J. R. Smith's *Tree Crops: A Permanent Agriculture*, 2nd ed. (New York: Devin-Adair Co., 1950). See also John W. Jeffries, *Wartime America: The World War II Homefront* (Chicago: Ivan R. Dee, 1996), and Giles Slade's *Made to Break: Technology and Planned Obsolescence in America* (Cambridge, MA: Harvard University Press, 2006). Churchill's quotation on the "U-boat peril" comes from Winston Churchill, *Their Finest Hour* (New York: Harcourt, 1986 reprint), p. 529.

For the discussion of the FBI's pursuit of saboteurs and the Duquesne spy ring, the FBI's Freedom of Information file on Frederick Duquesne (subtitle: "Interesting Case Write-up") lives up to its title (279 pp.). Duquesne is a secondary figure in Clint Richmond's book *Fetch the Devil: The Sierra Diablo Murders and Nazi Espionage in America* (Lebanon, NH: ForeEdge, 2014) but that book provides valuable insights, supported by *New York Times* articles of the period, especially "29 Suspects Show Variety of Talent," June 30, 1941. I had the benefit of the FBI's sabotage file on the Crown Cork and Seal fire of September 1940, which was declassified after my Freedom of Information Act (FOIA) request: FBI file 98-37, dated Nov. 3, 1940, made at Baltimore, MD.

## CHAPTER TWO. THE MARSAS RETURN TO SPAIN

The narrative of the Marsa family and Melchor Marsa's role in the cork industry developed through phone interviews with Gloria Marsa Meckel between August 2015 and April 2016. The story of the intelligence gathering by the Office of Strategic Services emerged from my discovery of cork industry contacts in the declassified OSS collection of the National Archives and Records Administration (NARA), starting from OSS memos about the prospect of Herman Ginsburg and/or Melchor Marsa collaborating with intelligence gathering (dated Oct. 2, 1942, and Oct. 7, 1942). A series of FOIA requests and interviews filled out a picture of Ginsburg's and Marsa's lives and work.

Marsa's experience with International Cork and the intellectual property suit with Crown Cork and Seal is covered in articles including the *Brooklyn Daily Eagle*, "500,000 Damage Suit Started in Boro Cork Co. War," Mar. 6, 1925. The *New York Times* reference to anarchist immigrants comes from "Large Number of Recent Crimes Due to Anarchism Explained by the Police as an Instance of Political 'Suggestion' Affecting an Entire Class..." (Mar. 15, 1908).

A valuable account of Americans and Spain during the 1930s is Adam Hochschild's *Spain in Our Hearts: Americans in the Spanish Civil War, 1936–1939* (New York: Houghton Mifflin, 2016). The account in "A Desperate Democracy Disregarded" is also very helpful (http://struggle.ws/spain/intervention.html). *El Negocio del Corcho en España Durante el Siglo XX*, by Francisco Manuel Parejo Moruno (Estudios de Historia Economica No. 57, Madrid, 2010) provides useful information on the cork industry there. The bombing of San Feliu de Guixols was reported by the *New York Times* ("Rebel Ships Shell Catalan Seaports," Feb. 24, 1938).

Readers interested in Portugal during the early 1940s should consult Neill Lochery's *Lisbon: War in the Shadows of the City of Light, 1939–1945* (New York: Public Affairs, 2011); *The Lisbon Route: Entry and Escape in Nazi Europe*, by Ronald Weber (New York: Ivan R. Dee, 2011), and, for the European refugee migration during the war, *European Junction*, by Hugh Muir (London: George G. Harrap, 1942). James M. Anderson's *The History of Portugal* is also helpful (Westport, CT: Greenwood Press, 2000). A novel about Lisbon during the period by Robert Wilson, titled *The Company of Strangers* (New York: Harcourt, 2001) also offers atmospherics. For contemporary journalists' accounts of Portugal, I drew especially from "Portugal and Her Dictator, Antonio Salazar" (*New York Times*, May 21, 1939), "Turbulent Gateway of a Europe on Fire" (*New York Times*, Mar. 23, 1941), and "Portugal, Europe's Refugee Capital, Fears Nazi Thrust" (*Washington Post*, Mar. 16, 1941).

The economic and trade situation of Portugal during the war was examined at the London Conference on Nazi Gold in 1997; see "Portugal and the Nazi Gold: The 'Lisbon Connection' in the Sales of Looted Gold by the Third Reich," by Antonio Louça and Ansgar Schäfer (*Yad Vashem Studies* 27 (1999): 105–123, Shoah Resource Center, 1999). A contemporary report relating to the cork trade is "Allies Act to Block Nazi Gold in Trade" (*New York Times*, Feb. 23, 1944).

## CHAPTER THREE. THE DICARAS IN A BIND

The narrative of Frank DiCara's family and his involvement with the cork industry I owe to his generosity. I might not have met DiCara if not for a fellow writer, Rafael Alvarez, who put me in touch with him. Complementing DiCara's detailed remembrances of life and factory work during the war are three main references: Richard R. Lingeman, *Don't You Know There's a War On?: The American Home Front, 1941–1945* (New York: Putnam, 1970); Mark Reutter, *Sparrows Point: Making Steel . . . The Rise and Ruin of American Industrial Might* (New York: Touchstone, 1989); and A. J. Baime's *The Arsenal of Democracy: FDR, Detroit, and an Epic Quest to Arm an America at War* (cited above). Contemporary reports include "Big Changes in Store for American Business" (*New York Times*, Nov. 23, 1941). (Unlike its namesake in Italy, Highlandtown's Our Lady of Pompei uses only one "i.")

Four valuable references on the predicament of Italian Americans during the war are: Lawrence DiStasi, ed., *Una Storia Segreta: The Secret History of Italian American Evacuation and Internment during World War II* (Berkeley, CA: Heyday Books, 2001); Stephen Fox, *Uncivil Liberties: Italian Americans Under Siege during World War II* (Parkland, FL: Universal Publishers, 2000), which updates Fox's earlier book, *The Unknown Internment: An Oral History of the Relocation of Italian Americans during World War II* (Boston: Twayne Publishers, 1990); and Salvatore J. LaGumina, *The Humble and the Heroic: Wartime Italian Americans* (Youngstown, NY: Cambria Press, 2006). Complementing these are two more recent books: Jan Jarboe Russell, *The Train to Crystal City: FDR's Secret Prisoner Exchange Program and America's Only Family Internment Camp during World War II* (New York: Scribner, 2015), and Lawrence W. DiStasi, *Branded: How Italian Immigrants Became "Enemies" during World War II* (Bolinas, CA: Sanniti Publications, 2016). DiStasi's books are the principal sources for Pascale DeCicco's ordeal. Thomas A. Guglielmo pointed me to Paul Campisi's master's thesis, "The Adjustment of Italian-Americans to the War Crisis," submitted to the Faculty of the Division of the Social Sciences, University of Chicago, December 1942. Guglielmo's book, *White on Arrival: Italians, Race, Color, and Power in Chicago, 1890–1945* (New York: Oxford University Press, 2004) provides insightful historical context on how the group's status changed.

The account of the war's effects on immigrant families in Maryland ("Rising wartime fears...") is informed by a number of articles from the *Baltimore Sun* including: "Judge

Denies Citizenship to 34 Aliens," (Mar. 9, 1941; on the arrest and prosecution of Michael Etzel: "Cullen Holds Plane Worker for Sabotage" (Oct. 30, 1941); and on his sentencing, "15-Year Term Given Etzel for Sabotage" (Nov. 18, 1941). *New York Times* coverage of the case included: "Painter Confesses to Plane Sabotage" (Oct. 30, 1941); "Plane Worker Is Held: Etzel Bound Over to Grand Jury on Baltimore Sabotage Charge" (Nov. 4, 1941); "Indicted as Saboteur" (Nov. 5, 1941); and "Saboteur Gets 15 Years" (Nov. 18, 1941). Newspapers also reported on the harbor sabotage by Italian ships' crews: "Coast Guard Lists Damages to Ships" (*New York Times*, Apr. 2, 1941) and "31 Italians Sentenced" (*New York Times*, July 18, 1941).

I retrieved the confidential OSS report on Highlandtown by agent L. R. Taylor, OSS Foreign Nationalities Branch, in Record Group (RG) 226, microfilm titled, "Samples of Italian Opinion in Baltimore," Apr. 10, 1944, at the National Archives at College Park, College Park, MD. The OSS research on the America First movement and its financial ties to Nazi Germany is described in Donald Downes's book *The Scarlet Thread* (London: Hunt, Barnard, 1953), pp. 61–63.

The rise of unions at Sparrows Point and the role of black workers there is described in the *Labor Herald*, a Baltimore-published weekly of the period, and Mark Reutter's book, *Sparrows Point: Making Steel... The Rise and Ruin of American Industrial Might*, cited above. For an insightful history of factories in general and information on the Glenn L. Martin factory in particular, see Joshua B. Freeman's *Behemoth: A History of the Factory and the Making of the Modern World* (New York: W. W. Norton, 2018), pp. 231-232.

Descriptions of gambling in Baltimore came from the 1940 *WPA Guide to Maryland: The Old-Line State* (p. 204) and reports in the *Baltimore Sun*: "Gambling and the Law: Is the Law an Ass?" (Oct. 21, 1939), "The Week's News" (July 18, 1943), and "Gored by Bull" (Aug. 1, 1943), along with my interviews with Frank DiCara, November 2013; February, May, and November 2014; May and October 2015; April and June 2016.

The Coast Guard volunteer patrols were described in "Baltimore: War Plants Are Heavily Guarded and Waterfront Is Restricted" (*New York Times*, Apr. 6, 1942) and "Women's Division of Port Security Force Growing" (*New York Times*, July 23, 1943). The Baltimore shipyard welder's arrest comes from "Welder Accused of Sabotage," *New York Times*, Mar. 23, 1943. The Boyd Stalnaker sabotage case was covered by the *Baltimore Sun*, especially in "War Worker Is Accused of Sabotage" (Apr. 12, 1942) and "Hose Damage Draws Term" (May 20, 1942).

*The Burning Shore: How Hitler's U-Boats Brought World War II to America*, by Ed Offley (New York: Basic Books, 2014) describes German U-boat activity in the Chesapeake Bay and off the Atlantic Coast. Regarding the slow progress to targets ("the country was supposed to have sent 5,300 fighter planes"), see Baime, *The Arsenal of Democracy*, pp. 131–134.

For more on how the war drew American workers into new relationships with the government and economy, see *American Workers, American Unions*, 3rd ed., by Robert H. Zieger and Gilbert J. Gall (Baltimore: Johns Hopkins University Press, 2002), p. 118. Baime describes Roosevelt's visit to the Ford bomber factory at Willow Run on pp. 169–173.

## CHAPTER FOUR. THE MCMANUS CORK PROJECT

Descriptions of the 1941 Santa Anita Handicap come from "Bay View, an Unknown, Wins at Santa Anita," by Russ Newland, *The Day*, Mar. 3, 1941, and a Universal newsreel of the race: https://www.youtube.com/watch?v=-2y_NMt5Lso. The description of McManus Sr.'s bet and win comes from my interviews with Charles McManus Jr. McManus Sr.'s status among Maryland's highest-paid executives is covered in the *North Adams Transcript*, front-page story on May 15, 1939, "Highest Salary for 1938 Was Paid George W. Hill," and his 1943 salary in the *Baltimore Sun*, Dec. 31, 1945, "5 Marylanders Listed Among High Earners."

"You wanna drive? You drive" comes from my interview with Charles McManus Jr., in March 2006. The story of McManus's idiosyncratic shorthand is also from that interview.

Portuguese cork suppliers' views come from *Boletim da Junta Nacional Corticeiras* no. 32 (June 1941): 4–5. Cork joins list of strategic materials: "OPM to Control Cork: Order Puts Stocks in Reserve, Subject to Defense Needs," *New York Times*, June 3, 1941, and "Cork in New Bottleneck," *New York Times*, July 7, 1941.

The details of Fritz Duquesne's shortwave communications from Long Island comes from the FBI FOIA file on Frederick Duquesne. File 8: https://archive.org/details/Duquesne.

For the FTC filing against Crown Cork, see "Bottle Cap Field Cited on Monopoly," *New York Times*, Oct. 4, 1941. The other factory fire: "Baltimore Defense Plant Burns," *New York Times*, Feb. 16, 1941. The Commerce Department report and the item on "the men from the Third Reich have been strongly attracting sellers" appeared in "Cork Goes to War," *Foreign Commerce Weekly*, Oct. 18, 1941.

The origins of the McManus Cork Project come from various sources by Giles B. Cooke.

Woodbridge Metcalf's background is reported in Woodbridge Metcalf, *Extension Forester, 1926–1956* (Berkeley: University of California, Bancroft Library, 1968), and many details come from Metcalf's field notebook and datebooks in the Woodbridge Metcalf papers, BANC MSS C-B 1018, Carton 21 and 22. Courtesy of the Bancroft Library, University of California, Berkeley.

For accounts of Italian Americans in the Bay Area, see Lawrence DiStasi, ed., *Una Storia Segreta* (cited above) on the relocation of families and suicides resulting, pp. 20 and 108, and Salvatore J. LaGumina's book, *The Humble and the Heroic: Wartime Italian Americans* (also cited above), p. 103. This section also references "Pays to Aliens under Study: Plans for Compensating Evacuated Groups Being Drafted on Coast," *Baltimore Sun*, Apr. 2, 1942.

The declassified memorandum of the December 30, 1941, meeting of Crown Cork and Seal executives with War Department officials is in the National Archives (with Jan. 6, 1942, transmittal slip from F. H. Otto to the undersecretary of war), along with David Stone's sketch of floor plans of the space to be devoted to defense work.

John McGrain, Baltimore County historian, generously answered my questions on his participation as a boy in the cork-growing campaign and other wartime recycling efforts, as well as Baltimore industrial responses to the war, in interviews from 2006 and email exchanges through 2016.

### CHAPTER FIVE. SERVING THE CROWN IN WARTIME PORTUGAL

For the account of the Marsa family in Portugal, I'm indebted to Gloria Marsa Meckel and her patience in phone interviews with me, from August 2015 through April 2016. For historical accounts of Salazar's Portugal, my main sources are: Neill Lochery, *Lisbon: War in the Shadows of the City of Light, 1939–1945* (cited above); Peter Fryer and Patricia McGowan Pinheiro's *Oldest Ally: A Portrait of Salazar's Portugal* (London: Dennis Dobson, 1961); and Douglas L. Wheeler's *Historical Dictionary of Portugal* (Metuchen, NJ: Scarecrow Press, 1993). Dr. Wheeler, professor emeritus of history, University of New Hampshire, described the pattern of the PVDE's treatment of internationals via email (email exchange with David Taylor, June 23–24, 2016), in "In the Service of Order: The Portuguese Political Police and the British, German and Spanish Intelligence, 1932–1945," *Journal of Contemporary History* 18(1) (1983): 1–25; and in Douglas L. Wheeler, "A Little World War II: Lisbon, 1939–1945" (unpublished manuscript, 2016).

Gloria Marsa Meckel described her father's assistance to the Leenaards family in August 1940 and the dinner in Lisbon in our April 4, 2016, interview. Other accounts of the period include Craig Thompson, "Portugal, Europe's Refugee Capital, Fears Nazi Thrust," *Washington Post*, Mar. 16, 1941.

Of the many accounts of the America First movement, this book draws on Baime's *The Arsenal of Democracy*.

The *Sines* cargo fire is described in "Part of Cork Cargo Burns," *New York Times*, Sept. 30, 1941.

Three important accounts of the FBI's arrest of the Fritz Duquesne spy ring are in: Peter Duffy, *Double Agent: The First Hero of World War II and How the FBI Outwitted and Destroyed a Nazi Spy Ring* (New York: Scribner, 2014); David Alan Johnson, *Betrayal: The True Story of J. Edgar Hoover and the Nazi Saboteurs Captured during WWII* (New York: Hippocrene Books, 2007); and William Breuer, *Nazi Spies in America: Hitler's Undercover War* (New York: St. Martin's, 1989). The FBI newsreel cited is "Battle of the United States," *Army-Navy Screen* magazine #42, Washington, DC, 1944.

Shipping and the *Pero de Alenquer* held something closer to celebrity status in the transatlantic heyday, judging by coverage in the *New York Times*: "New Service to Lisbon" (Aug. 17, 1940); "Portuguese Vessel Delayed by Storms" (Dec. 31, 1940); "12 Passengers Arrive on Ship from Lisbon" (June 8, 1941); "Cork from Portugal to Float U.S. Defense Program" (Aug. 8, 1941); and "Portuguese Ship Arrives: Is First Neutral Vessel Here Since U.S. Entered War" (Dec. 19, 1941). Two declassified OSS memos of interviews with the captain of the Pero de Allenguez (*sic*), are found in the OSS records RG 226, COI/OSS Central Files, Container 317, the National Archives at College Park, College Park, MD. The first is dated July 8, 1942, and a second, handwritten memo is dated July 9, 1942. Donald Downes's quotations about the OSS ship observer program come from *The Scarlet Thread* (cited above), pp. 75–76.

The account of José Robert is detailed in the OSS records at the National Archives,

RG 226, COI/OSS Central Files, Container 317; the undated two-page memo on the SS *Thetis* is filed between June and July 1942. The next two-page memo dated July 20, 1942, begins, "It seems sure that from now on, all our boats are going to arrive in Baltimore." Other memoranda in that container describe other ships and their crews, some docked in Baltimore; some of the memos are addressed to "A.D.," probably Allen Dulles.

Among the most useful histories of the OSS I found are: R. Harris Smith, *OSS: The Secret History of America's First Central Intelligence Agency* (Berkeley: University of California Press, 1972); George C. Chalou, ed., *The Secrets War: The Office of Strategic Services in World War II* (Washington, DC: National Archives & Record Service, 1992); Bradley F. Smith, *The Shadow Warriors: O.S.S. and the Origins of the C.I.A.* (New York: Basic Books, 1983); Lt. Col. Corey Ford and Maj. Alastair MacBain, *Cloak and Dagger: The Secret Story of OSS* (New York: Random House, 1946); and Robin Winks, *Cloak & Gown: Scholars in the Secret War, 1939–1961* (New Haven: Yale University Press, 1996). An overview of the recruitment process and profile of OSS informants and agents appears in the book by OSS Assessment Staff, *Assessment of Men: Selection of Personnel for the Office of Strategic Services* (New York: Rinehart & Co., 1948). More information on the agency's organization appears in *The Overseas Targets: War Report of the OSS*, vol. 2.

The OSS memos and communiqués I quoted about the *Thetis* and *San Miguel* were consulted August 2015, in RG 226, Entry 14—R&A Reports, Portugal, Records of the Office of Strategic Services, National Archives at College Park, College Park, MD.

Donald Downes describes his early career with British Naval Intelligence in *The Scarlet Thread*, pp. 10 and 60. His declassified memo to Allen Dulles, dated July 11, 1942, describing his view of stoking rebellions in Europe, is in the OSS collection, Allen Dulles folder. The account of the Italian community in Springfield, MA, is in a file headed "Springfield, Mass.—List of known Italian Fascist Workers—compiled and cross-checked," in the Donald Downes file in the Records of the Office of Strategic Services, RG 226, National Archives at College Park, College Park, MD, 2 pages (n.d.).

The fullest accounts of Operation Pastorius and the Nazi saboteurs episode in America during the summer of 1942 are in David Alan Johnson, *Betrayal: The True Story of J. Edgar Hoover and the Nazi Saboteurs Captured during WWII*, and William Breuer, *Nazi Spies in America: Hitler's Undercover War*, both cited above. The news account cited is "Nazi Saboteurs Face Stern Army Justice" (*New York Times*, July 4, 1942).

The declassified memos by Robert Ullman about his meetings with Herman Ginsburg, dated Oct. 2, 1942, and Oct. 7, 1942, are found in RG 226, Records of the Office of Strategic Services, National Archives at College Park, College Park, MD. The Crown coworker's description of Ginsburg comes from W. Michael Blumenthal, *From Exile to Washington: A Memoir of Leadership in the Twentieth Century* (New York: Overlook Press, 2013). The description is rounded out by the recollections of Ginsburg's niece, Connie Robinson (phone call with author, Apr. 6, 2015) and the declassified FBI file on Herman Ginsburg, dated Mar. 4, 1955, and Mar. 30, 1955, obtained through a FOIA request.

The *Brooklyn Daily Eagle* article about Melchor Marsa's transatlantic passage on the Pan American Clipper is dated Aug. 23, 1941. "Nazis Order Ships Routed to Baltimore," *Washington Post*, July 10, 1942.

The account of the Casablanca conference comes from Rick Atkinson, *An Army at*

*Dawn: The War in North Africa, 1942–1943* (New York: Henry Holt, 2002). I also consulted Henry Gerard Phillips's *Sedjenane: The Pay-off Battle* (Penn Valley, CA: Henry Gerard Phillips, 1993).

Armstrong Cork employee David E. Sanderson appears in the OSS Personnel Files, RG 226, Entry 224, Records of the Office of Strategic Services, National Archives at College Park, College Park, MD. He is first mentioned in the Oct. 2, 1942, memo by Robert Ullman about his meeting with Herman Ginsburg, where Ullman states, "Through Mr. Dudley Armstrong I heard of David Sanderson, an able young cork man, who wants to go into the Navy . . . if [Armstrong is] persuaded this was a useful mission, Sanderson might be a choice for C.C. & S. International." Sanderson's background and mission are described in a memo from F. L. Mayer to Francis M. Barker dated Aug. 17, 1943 (2 pages).

The figures on Crown Cork's values in 1941 come from *Crown Cork & Seal Company, Inc. Annual Report*, year ended Dec. 31, 1941 (Baltimore: Crown Cork & Seal, 1942).

The sinking of the *Molly Pitcher* is recounted in Robert M. Browning, *U.S. Merchant Vessel Casualties of World War II* (Annapolis: Naval Institute Press, 1996), 303; and "Molly Pitcher: American Steam Merchant" on the website Ships Hit By U-Boats (http://uboat.net/allies/merchants/2808.html).

## CHAPTER SIX. AMONG THE SPIES IN LISBON

The epigraph from Donald Downes's memo to OSS director William Donovan comes from Douglas Waller, *Wild Bill Donovan: The Spymaster Who Created the OSS and Modern American Espionage* (New York: Free Press, 2011), p. 111.

Ronald Weber's *The Lisbon Route: Entry and Escape in Nazi Europe* (cited above) is the source of Graham Greene's observations, p. 273, as well as the *Saturday Evening Post* quotation, p. 233.

Melchor Marsa's meeting with OSS agent Van Halsey is recorded in Halsey's internal OSS memo to Lloyd Hyde dated June 15, 1943, item 19732B in RG 226, Entry 214. OSS goals for Spain and Portugal are described in *The Overseas Targets: War Report of the OSS*, vol. 2, p. 31.

J. Edgar Hoover's letters to William Donovan regarding commerce with Portugal are noted on index cards of OSS internal communiqués including: Portugal's tinplate output (dated Jan. 15, 1942, item 9898 C), smuggled platinum (dated Jan. 31, 1942, item 10829 C), tungsten (dated Feb. 21, 1942, item 12752 C), and hides from Brazil (dated Mar. 27, 1942, item 14195 C). For the German Press Summaries, see British Press Office, Lisbon, "German Press Summaries on Portugal and Spain, 1943" (not published), Library of Congress, Washington, DC.

The July 1943 Salazar government response to labor unrest: OSS communiqué dated July 31, 1943 (item 41271 R); rise in prices and public dissatisfaction (item 41296 S, dated Aug. 9, 1943); "Portugal might even enter the war . . ." (item 44393C, dated 8/ /43[*sic*]). Gardner McPherson's records in the OSS files detail his pay and dates in Lisbon; his interests are described in "Gardner B. MacPherson," *Washington Post*, Jan. 30, 1993.

Lisbon as the only port for Crown Cork and Seal exports to the United States is noted by various sources including my interviews with Charles McManus Jr., 2006. The role of

Joaqium Vieira Natividade in the Portuguese cork industry is well documented in Ignacio García Pereda, "The Junta National da Cortiça (Cork Portuguese Board, 1936–1972) and the Transference of Technology. Cork Products Research is from the Portuguese 'Estado Novo' Regime," *XXXI Conference of the Portuguese Economic and Social History Association*, Faculdade de Economica da Universidade de Coimbra, 2011, http://www4.fe.uc.pt/aphes31/index_en.html.

Neill Lochery describes Setúbal's port and wartime intrigue in *Lisbon: War in the Shadows of the City of Light, 1939–1945*; further details on locations of the Crown Cork factory and nearby facilities came from the Divisão da Cultura e Património, Ecomuseu Municipal Serviço Educativo Núcleo da Mundet. Gloria Marsa's life in Brooklyn comes from interviews with her, August 2015 and April 2016. Noah Isenberg details the links between the movie *Casablanca* and Lisbon in his book, *We'll Always Have Casablanca: The Life, Legend, and Afterlife of Hollywood's Most Beloved Movie* (New York: W. W. Norton, 2017), pp. 6 and 129. Douglas L. Wheeler described the state of the Spain-Portugal border and smuggling in my email exchange with him June 23–24, 2016; Stanley G. Payne, professor emeritus, History Department, University of Wisconsin, shared information about the border area also, in our email exchange May 16–June 2, 2016. Melchor Marsa described his return from Argentina to his daughter Gloria, who relayed it in my interviews with her.

OSS communiqués more directly on cork and tungsten were: Virgilio Calixto Pires (item 65048S dated Feb. 10, 1944), Lomelino Lira (item 73539S dated Mar. 27, 1944), factory of Horacio Coelho (item 62229S dated Nov. 25, 1943), and the railway quay at Vila Pereira (dated Aug. 1, 1944).

See "Allies Act to Block Nazi Gold in Trade," *New York Times*, Feb. 23, 1944. See also Antonio Louça, and Ansgar Schäfer, "Portugal and the Nazi Gold: The 'Lisbon Connection' in the Sales of Looted Gold by the Third Reich," *Yad Vashem Studies* 27 (1999): 105–123. The OSS communiqués on Kuehne & Nagel are in the NARA RG 226 collection, dated June 10, 1944, item 75758.

Operation Banana is described in Smith, *OSS: The Secret History of America's First Central Intelligence Agency*, pp. 79–82; in Downes, *The Scarlet Thread* (cited above), pp. 99–121; and in Robin Winks, *Cloak & Gown: Scholars in the Secret War, 1939–1961*, pp. 198–204.

Forester Palmer Stockwell described his debts to his hosts in Spain and Portugal in "The Culture of Cork Oak in Spain," *Economic Botany* 1(4) (Oct.–Dec. 1947): 381–388. Other aspects of the cork harvests of that period and the *montado* come from my interview with Isidoro Rodero Jímenez at Correia Cork, in Azaruja, Portugal, Dec. 22, 2016, and from personal communication with forest economist Amélia Branco, University of Lisbon, Dec. 23, 2016. Portuguese cork suppliers' views of the McManus Cork Project of the time come from *Boletim da Junta Nacional Corticeiras*, Nov. 1943 issue.

Frances Perkins's role in international affairs is described by Kirstin Downey in her biography, *The Woman behind the New Deal: The Life and Legacy of Frances Perkins—Social Security, Unemployment Insurance, and the Minimum Wage* (New York: Anchor Books, 2009), and in Downey's phone conversation with me on Oct. 17, 2016. Downey noted that Perkins "did a lot more in Immigration than is in any public record." Although by 1940 Perkins had no official duties with Immigration, she had been concerned by the actions of Nazi Germany since 1933, and Downey reported she found many informal reports of how Perkins intervened on behalf of people escaping Europe.

The OSS memo on visits to the engineer aboard *San Thomé* is in the NARA RG 226 collection, item 65940S, dated Feb. 22, 1944. The OSS communiqué on the arrest of stevedores in a sabotage attempt on the *Margaret Johnson* is 57752S, dated Feb. 9, 1944.

The example of German officials' leveraging sale of an automobile for intelligence obligation comes from Neill Lochery, *Lisbon: War in the Shadows of the City of Light, 1939–1945*, pp. 144–145. OSS Lisbon office politics come up in RG 226, Lisbon Field reports, memorandum dated Mar. 26, 1945, from Lois Lombard addressed to Chief, SI, 2 pages.

Marsa's return on the *Henry Hadley* is documented in the ship manifest on Ancestry .com: https://www.ancestry.com/interactive/Print/7488/NYT715_7037-0066, accessed May 8, 2017.

## CHAPTER 7. FROM THE FACTORY TO THE FRONT

The summary of US industrial production and labor strikes in 1943 comes from "New Records Made by Industry in '43," *New York Times*, Jan. 2, 1944. The flight crew assessment of improvements in the B-26 performance comes from "Transcript of an Oral History Interview with John B. Hodgson, Gunner, Air Force, World War II," Wisconsin Veterans Museum Research Center, 2000, pp. 14–15. Hodgson served in the Army Air Corps from 1943 to 1945.

Baltimore's racial tensions and the ACLU report cited appear in "Race Problem Called Acute," *Baltimore Sun*, Oct. 27, 1943. The number of Maryland men in the war comes from "171,500 State Men in Armed Forces," *Baltimore Sun*, May 1, 1944. Frank DiCara's deployment date and unit come from interviews with him in November 2013; February, May, and November 2014; May and October 2015; April and June 2016; and are confirmed by his US military record.

Woodbridge Metcalf's letter from Ralph Waltz, dated Mar. 12, 1943, is item F38, Cart 1, in the Woodbridge Metcalf papers, BANC MSS C-13 1018; courtesy of the Bancroft Library, University of California, Berkeley. Other details of Metcalf's schedule in late 1944/early 1945 come from his datebooks in that collection.

The wartime ads of Armstrong Cork appeared in *Saturday Evening Post*. For the role of the B-26 in D-day, see Steven Ambrose, *D-Day, June 6, 1944: The Climactic Battle of World War II* (New York: Simon & Schuster, 1994), and for the plane's record throughout the war and the 1946 assessment, see *Arizona Warplanes*, by Harold A. Skaarup, updated ed. (New York: iUniverse Inc., 2010), p. 98.

The account of Berkeley's support for the USS *City of Berkeley* comes from Margot Lind, "Down to the Seas in Berkeley's Ships," *Newsletter of the Berkeley Historical Society* 27(1) (Spring 2009): 1ff. For Philip K. Dick's early employment, see Lawrence Sutin, *Divine Invasions: A Life of Philip K. Dick* (Boston: Da Capo, 2005). For the story of Pittsburg's Italian residents and Nino Guttadauro, see Lawrence DiStasi, *Una Storia Segreta: The Secret History of Italian American Evacuation and Internment during World War II* (cited above). More about Executive Order 9066 is in Stephen Fox, *Uncivil Liberties: Italian Americans Under Siege during World War II*, and Salvatore J. LaGumina, *The Humble and the Heroic: Wartime Italian Americans* (both cited above).

## CHAPTER EIGHT. POLITICS AND GASOLINE

The Woodbridge Metcalf epigraph comes from *Extension Forester, 1926–1956* (cited above). The accounts of Arbor Day celebrations with the McManus Cork Project come from various festival programs and reports by Giles B. Cooke, in his collected papers at the Swem Library at the College of William and Mary, Williamsburg, VA. These include: "Alabama Arbor Day Celebration Tuesday, February 29, 1944" (reprint from April 1944 issue of *The Crown*); "Arbor Day in Alabama" by Giles B. Cooke and Clifton F. Schmidt Jr.; "Governor Laney Plants Cork Tree" by Giles B. Cooke; "California Arbor Day in California 1944"; "Florida Cork Tree Planting and Dedication Tuesday, December 12, 1944" (reprint from Feb. 1945 *The Crown*); "Cork Oak Planted by Governor Holland" by Giles B. Cooke and Clifton F. Schmidt Jr.; "Georgia Arbor Day Ceremony Friday, December 1, 1944" (reprint from Jan. 1945 *The Crown*); "Georgia Celebrates Arbor Day" by Giles B. Cooke and Clifton F. Schmidt Jr.; "Louisiana Arbor Day Ceremony Friday, January 26, 1945" (reprint from Jan. 1945 *The Crown*); "Louisiana Observes Arbor Day" by Giles B. Cooke and Clifton F. Schmidt Jr.; "Maryland Arbor Day Celebration Friday, April 5, 1946" (reprint from June 1946 *The Crown*); "Maryland Observes Arbor Day" by Giles B. Cooke; "Mississippi Planting and Dedication of a Cork Oak Tree Wednesday, March 15, 1944" (reprint from May 1944 *The Crown*); "Governor Bailey Plants Cork Oak" by Giles B. Cooke and Clifton F. Schmidt Jr.; "North Carolina Arbor Day Ceremony Friday, March 21, 1947" (reprint from May 1947 *The Crown*); "Arbor Day in North Carolina" by Giles B. Cooke; "Cork Tree Planted by Governor McCord" by Giles B. Cooke; "South Carolina Arbor Day Friday, December 3, 1943" (reprint from Dec. 1943 *The Crown*); "South Carolina Plants Cork" by Giles B. Cooke and Clifton F. Schmidt Jr. (reprint from Jan. 1944 *The Crown*), and "Governor Johnson Plants Cork Tree."

The work at Armstrong Cork comes from "War Production at Armstrong Cork Co." in *Lancaster County Historical Society Journal* 100(4) (Winter 1998); and William A. Mehler Jr., *Let the Buyer Have Faith: The Armstrong Story*, cited above.

Besides the daybook of Woodbridge Metcalf's cross-country trip in January–February 1945, see "U.C. Professor Helps Cork Culture in U.S.," *Los Angeles Times* (Carton 22, Folder 6) in the Woodbridge Metcalf papers, BANC MSS C-B 1018, courtesy of the Bancroft Library, University of California, Berkeley. See also "Cork Oak Production Fostered in Arizona," *Arizona Republic*, Feb. 4, 1945; "Army Ammunition Explosion Rocks Southwest Area," *El Paso Herald-Post*, July 16, 1945; and "Million Cork Acorns Sown in 10 States," *Washington Post*, Feb. 21, 1945.

For how Crown Cork and Seal performed through the war, see Crown Cork and Seal, Annual Report 1944, Baltimore, MD; and my interviews with Charles McManus Jr., January and March 2006. The Waldemar Kaempffert article on plastics is "Balanced Survey of the Plastic Age," *New York Times* book review, July 23, 1944.

Charles McManus Jr. interviews in 2006 are the source for the postwar trip to Crown Cork offices in Europe and his reflections on the McManus Cork Project's conclusion.

For General George Marshall's assessment of the war, see "Marshall Tells How Near Defeat Allies Were in 'Black Days of '42,'" *Baltimore Sun*, Oct. 10, 1945. The two Giles B. Cooke publications are "Growing Cork at Home," *Journal of Chemical Education* 22(8) (Aug. 1945): 367, and "Tree with a Future," *Science Illustrated* 6(7) (Sept. 1945): 51. The article on the Hercules Powder factory blast is W. Barton Leach, "Preventing Sabotage Called Possible and Necessary," *New York Times*, June 8, 1941. See also "Bottle Cap Field Cited on Monopoly: FTC Complaint Names Crown Makers Group and Fourteen Manufacturers," *New York Times*, Oct. 4, 1941. "Charles M'Manus, Cork Firm's Head: Chairman of Board of Crown Cork and Seal Co., Inventor of a Bottle Seal, Dies," appeared in the *New York Times*, June 4, 1946.

## CHAPTER NINE. COLD NEW WORLD

In addition to Gloria Marsa Meckel's remembrances from our interviews, this chapter draws on W. Michael Blumenthal, *From Exile to Washington: A Memoir of Leadership in the Twentieth Century* (cited above) and my two phone interviews with Secretary Blumenthal, August 24, 2015, and February 24, 2016. The account of Herman Ginsburg also comes from my interviews with his niece, Connie Robinson, Apr. 6, 2015; and from the FBI case file 100-HQ-417406, dated March–June, 1955. The memories of Ginsburg by Charles McManus Jr. come from my phone interviews with his son, David McManus.

The CIA report cited is "Cork Production and Trade with Particular Emphasis on the Soviet Bloc," Intelligence Memorandum No. 361 (CIA/RR IM-361), Nov. 26, 1951, released by CIA Historical Review Program 1999, 12 pages. The Soviet minister quoted comes from "'Down with War!' Bulganin Toasts: But Soviet Is Ready to Fight, He Says at Party Attended by Western Officials," *New York Times*, Aug. 18, 1954. On the Cold War trade, see "French Lift Red Trade: 18-Month Soviet Pact Calls for 60 Per Cent Increase," *New York Times*, Nov. 11, 1954.

The historical context of America's growing share of Mid-East oil in the 1940s comes from Michael Tanzer, *The Energy Crisis: World Struggle for Power and Wealth* (New York: Monthly Review Press, 1974), pp. 14–15.

## CHAPTER TEN. WAKING UP IN AMERICA

The reported figure of soldiers serving in the Pacific who reported suffering from malaria was between 60 and 65 percent, according to "Malaria in World War II" (Army Heritage Center Foundation, Carlisle, PA, https://www.armyheritage.org/75-information/soldier-stories/292-malaria-in-world-war-ii).

The situation that many veterans faced returning from World War II is well told in Thomas Childers, *Soldier from the War Returning: The Greatest Generation's Troubled Homecoming from World War II* (New York: Houghton Mifflin Harcourt, 2009). Here again I owe a debt to Frank DiCara for the account he shared in our interviews, November 2013; February, May, and November 2014; May and October 2015; April and June 2016. The results of the July 1944 Gallup poll about Americans' expectations of postwar unemployment and US Labor Department forecasts appear in Samuel Greengard, "Fighting for Employment: Veterans in the '40s and Today," Workforce.com, Feb. 12, 2012, http://www.workforce.com/2012/02/22/fighting-for-employment-veterans-in-the-40s-and-today/.

The account of postwar shortages at Armstrong Cork comes from William A. Mehler Jr.'s *Let the Buyer Have Faith*, cited above.

Woodbridge Metcalf's assessment of the cork-growing project comes from *Extension Forester, 1926–1956*, cited above.

### EPILOGUE. TREASURY BALANCE

Figures on the US cork industry today come from the Cork Institute of America, Jerry Manton. The description of the cork ecosystem now is informed by Alastair Bland, "Cork Trees: Soft-Skinned Monarchs of the Mediterranean," *Smithsonian* magazine, June 28, 2012. My email exchange with Glenda Pittman Owens about her father's planting of cork trees took place between June 24, 2016, and July 24, 2016. My phone conversations with Sarah Corbin about her and her brother's plantings as part of the McManus Cork Project were in August 2016.

Information on rare earth minerals comes from various sources including Virginia Tech, "Researchers Seek Ways to Extract Rare Earth Minerals from Coal," press release dated March 15, 2016, http://www.eurekalert.org/pub_releases/2016-03/vt-rsw031516.php.

# INDEX